P9-DMH-637

THE CLASS

THE CLASS

A Life-Changing Teacher, His World-Changing Kids,

and the Most Inventive Classroom in America

Heather Won Tesoriero

BALLANTINE BOOKS

NEW YORK

The Class is a work of nonfiction. Some names and identifying details have been changed.

Copyright © 2018 by Heather Won Tesoriero

All rights reserved.

Published in the United States by Ballantine Books, an imprint of Random House, a division of Penguin Random House LLC, New York.

BALLANTINE and the HOUSE colophon are registered trademarks of Penguin Random House LLC.

LIBRARY OF CONGRESS CATALOGING-IN-PUBLICATION DATA
Names: Tesoriero, Heather Won, author.
Title: The class : a life-changing teacher and his world-changing kids / Heather Won Tesoriero.
Description: New York : Ballantine Books, 2018.
Identifiers: LCCN 2018014535 (print) | LCCN 2018031851 (ebook) |
ISBN 9780399181856 (Ebook) | ISBN 9780399181849 (hardback : alk. paper)
Subjects: LCSH: Science—Study and teaching (Secondary)—Connecticut—
Greenwich. | Greenwich High School (Greenwich, Conn.) | Bramante, Andy. |
Science teachers—Connecticut—Greenwich. | High school students—Connecticut—
Greenwich. | Science fairs—United States. | Science projects—Competitions—
United States. | BISAC: SCIENCE / Study & Teaching. | EDUCATION /
Special Education / Gifted.
Classification: LCC Q183.3.C83 (ebook) | LCC Q183.3.C83 G747 2018 (print) |
DDC 507.1/273—dc23
LC record available at https://lccn.loc.gov/2018014535

Printed in the United States of America on acid-free paper

randomhousebooks.com

2 4 6 8 9 7 5 3 1

First Edition

Book design by Caroline Cunningham

For K, G + E,

who taught my heart everything it ever needed to know

and

in loving memory of my brother,

Matthew Robert Tesoriero

(1983–1991)

I cannot teach anybody anything. I can only make them think.

—Socrates

It's not the broken dreams that break us. It's the ones we don't dare to dream.

—Mr. Schue on *Glee*

Contents

Author's Note

- -

I spent the entire 2016–17 school year with Andy Bramante and his science research students. I commuted daily to Greenwich High School, minus occasional days off to write, and traveled to Washington, D.C., Los Angeles, and multiple locations in Connecticut to report on the science competitions.

I recorded more than seventy hours of interviews, conversations, and presentations. I also kept digital and analog notes from my time with Andy and his kids. It was impossible to be present for every scene in the book, so some have been reconstructed following extensive interviews.

There are several instances where names have been changed. In each case, I've explicitly stated when I've done so.

THE CLASS

The Science Fair Circuit

-- --

Each year, Andy Bramante's science research students compete in the following fairs/contests:

- **Siemens Competition in Math, Science and Technology**
  This contest was established in 1999 by the Siemens Foundation. Nearly two thousand students submit research papers. Five hundred semifinalists are named and compete live in six regional competitions, which culminate in a national competition every December. National finalists receive $25,000. A total of $500,000 in scholarship money is awarded, with the two top winners receiving $100,000 each. The final year of the competition was 2017.

- **Google Science Fair**
  Established in 2011, the Google Science Fair is a global fair where kids from all around the world submit their research online, as well as a short video explaining their work. Regional finalists are named,

followed by finalists who are invited out to Google's Silicon Valley headquarters to compete for the Grand Prize of $50,000 in scholarship money. (Google didn't hold the fair in 2017 but is planning to bring it back in the fall of 2018.)

- **Regeneron Science Talent Search (STS)**
  Created in 1942 by the Society for Science and the Public, STS is for high school seniors only and is considered to be one of the most prestigious and toughest science and math competitions. From its founding until 1997, STS was sponsored by Westinghouse. From 1998 to 2016, Intel sponsored the competition. In 2017, biopharmaceutical company Regeneron took over sponsorship and increased the prize amounts. Entrants submit a lengthy, rigorous application similar to a college application. About eighteen hundred seniors apply; three hundred are named scholars, or semifinalists; forty are later named finalists. The top forty seniors are invited to Washington, D.C., for a five-day competition. Scholars receive $2,000, with an additional $2,000 going to each scholar's school. The forty finalists receive a minimum of $25,000. At the event in Washington, the top ten winners receive prizes ranging from $40,000 to $250,000.

- **Connecticut STEM Fair (CT STEM)**
  Started in Westport, Connecticut, in 2001, this fair is for students in Fairfield County, Connecticut, and those at Amity Regional High School, located in Woodbridge, Connecticut. Roughly 250 students compete in the categories of health, physical, environmental, and behavioral science. They can compete with completed projects or proposals for yet-to-be-completed work. In addition to modest cash awards, CT STEM also awards scholarships to graduating seniors. As of 2017, three CT STEM winners were granted berths to attend Intel ISEF, the world's largest science fair. Starting in 2018, the fair has also been granted four spots at the GENIUS Olympiad at the State University of New York at Oswego.

- **Connecticut Science and Engineering Fair (CSEF)**
Founded in 1949, CSEF is the state's oldest and only statewide science fair. More than six hundred students compete in the fair. There are major life sciences and physical sciences categories and ten special categories that cover everything from mathematics to biotechnology. The fair and its 106 sponsors give more than $200,000 in awards and scholarships, as well as seven Intel ISEF slots.

- **Connecticut Junior Science and Humanities Symposium (JSHS)**
This competition is a collaborative effort with the research arm of the Department of Defense and is administered in cooperation with colleges and universities nationwide in an effort to promote scientific research. There are regional JSHS competitions held all over the United States and its territories. Winners advance to the national JSHS competition held each spring. CT-JSHS has been in existence for more than fifty years. Local competitors submit their research and a video discussing their work. Those who advance are selected to present posters of their work, or to deliver twelve-minute public presentations. Winners earn cash prizes ranging from $250 to $2,000. The top winner receives a four-year half-tuition scholarship to the University of Connecticut. The top five (first through fourth in orals and the first-place poster presenter) travel with all expenses paid to compete at the national JSHS competition, where winners receive $350 to $12,000.

- **Intel International Science and Engineering Fair (Intel ISEF)**
Intel ISEF is the world's largest precollege science competition, with nearly 1,800 kids competing from more than seventy-five countries, regions, and territories. It began as the National Science Fair in 1950 in Philadelphia and went international in 1958. Competitors must win Intel berths at local or regional affiliated fairs in order to attend ISEF, where $4 million in prizes are awarded, including one top award of $75,000 followed by two awards of $50,000.

- **International Sustainable World (Engineering Energy Environment) Project (ISWEEEP)**
Founded by Harmony Public Schools in Texas, ISWEEEP is a global science fair that draws four hundred kids from sixty countries. ISWEEEP has affiliated fairs around the world, but kids can also apply directly to compete. Held annually each spring in Houston, ISWEEEP awards $100,000 in prize money and scholarships. Winners ranging from bronze to Grand Awards receive $150 to $1,500. (In early 2018, ISWEEEP announced that it will not be holding the competition this year, but it hopes to return.)

- **Norwalk Science Fair**
This local fair is held each spring at Norwalk Community College in Norwalk, Connecticut. As many as eighty-five kids compete. Eight receive honorable mentions, and the top three winners receive $100 to $300.

- **Phenomenon: Science Innovation Fair**
Started in 2015 by the Bruce Museum in Greenwich, Connecticut, this fair is open to students in grades nine through twelve in Fairfield County, Connecticut, and Westchester County, New York. Members of the public are welcome to meet the top ten competitors and learn about their work. The first-prize winner receives $1,000, second takes home $500, and the winner of the People's Choice Award, voted on by the public, receives $250.

- **GENIUS Olympiad**
This international fair focuses on environmental issues. It was founded by the nonprofit group Terra Science and Education Foundation. It's held at the State University of New York at Oswego. Roughly five hundred students compete, vying for honorable mention, Bronze, Silver, Gold, Top Grand Gold, and special awards.

- **Davidson Fellows Scholarship**

  Founded in 1999, the Davidson Institute for Talent Development awards annual Fellow Scholarships of $10,000, $25,000, and $50,000 to high school students for their individual science projects. There is no live competition; work is judged based on the submission materials. Students are notified on or before July 15.

SUMMER

1

Andy

On July 18, 2016, at 2:22 A.M., Andy Bramante's cellphone buzzed with an incoming text. He rolled over in his bed, pulled his phone close to his face, and saw a message from Shobhita Sundaram.

"Heyy—guess who's a regional finalist for Google science fair??!" read the little text bubble.

Squinting down at his phone, Andy tapped out a reply. "Yaaaaaayyyy!!!! So happy for ya!"

Shobhi had spent her sophomore year in Andy's science research class at Greenwich High School building an algorithm that could help predict which breast cancer drugs would be most effective in killing tumors. She'd taught herself to code in middle school and was also a masterful public speaker, a crucial weapon in the science fair world. Between the knockout algorithm and her impressive ability to tout it, Andy thought Shobhita—aka Shobhi, the Shobhinator, and Shobhi-Wan Kenobi—was poised to floor the judges at the Connecticut Science and Engineering Fair (CSEF), the gateway to Intel ISEF, the mother of all science fairs. So it had seemed a shocking oversight of sorts when she failed to place at CSEF back in March. It had also infuriated Shobhi's mother, who'd wanted to send a scathing email to the judges.

Now, as he read the text that Shobhi was a regional finalist for the Google Science Fair, where thousands of kids from all over the world submitted a slew of ambitious projects, everything from fighting drought with fruit to detecting lung cancer with a breath test, a sense of redemption washed over Andy. He was thrilled that Shobhi got her due.

He fell back to sleep, but less than an hour later, his phone buzzed again. The 3:10 A.M. message was from William Yin, a rising senior and one of the most astounding kids Andy had ever had in his class.

"Hi Mr. B.," read the first text bubble.

The second said, "Shobi [sic] and I are GSF regional finalists!"

Ah, William. This was not so much a big surprise to Andy as continued validation for William's invention. He'd developed a blueprint for a Band-Aid-like sticker that could be applied to the neck to detect the presence of atherosclerosis, the plaque that can break off, form a clot, and kill you with a heart attack or stroke. The Google accolade was the latest glittery float in a seemingly endless parade of William's victories for his sticker.

Andy tapped out a congratulatory reply that involved a lot of exclamation points, trying to make his delight register through his foggy haze. William would always be one of Andy's most beloved, if crazy-making, students. There was a deep connection between the two, one born of brutally long days in the lab and the process of trying to work through the enigma that was William. But now Andy just wanted to get some shut-eye. He set his phone down on his night table, rolled over, and went back to sleep.

When he finally woke up for good that day, Andy texted back and forth to get some more reaction from the kids about the news. Shobhi was especially giddy. The night before, she'd set her alarm to wake up at two A.M. Eastern time, when Google released the names. Shobhi figured if she wasn't among this first cut of winners, she could just return to bed and sleep off her disappointment, rather than get up at a normal hour, learn she hadn't advanced, and spend her whole day being dogged by it.

Bleary-eyed, she woke up from her phone's alarm, opened her laptop, and logged on to the Google Science Fair site. And there on the page announcing the one hundred regional finalists were her photo and project. (Her region being the Americas.) The sixteen global finalists would be named in August. Shobhi texted Andy that her parents were "soo happy" and "I was too . . . once my heart rate had gone back to normal. I swear I got a heart attack when I saw my face on the page."

For the second year in a row, there was just one school and one teacher in the world with two Google regional finalists: Greenwich High and Andy Bramante.

Andy's science research class is unlike any other class at Greenwich High, a Connecticut public school behemoth with 2,650 kids. There is no curriculum, no tests, textbooks, or lectures. Students pitch individual projects that they work on throughout the entire school year with the goal of taking their discoveries and inventions out on the national and global science fair circuit. This is not the stuff of vinegar volcanoes and ant farms. Kids tackle problems like cancer, Parkinson's disease, HIV, heart disease, cheap water filtration, and carbon dioxide capture, sometimes making discoveries that elude adult scientists three times their age.

Andy Bramante arrived at GHS in 2005 as a chemistry teacher and picked up the research program the following year. In the years since, he has turned the program into a juggernaut, amassing one of the most impressive track records in America when it comes to kids competing in the science fair world. Year after year, his kids dominate the fair circuit and take home prizes in unprecedented numbers—all of which is particularly impressive because Andy is far from a career educator. He walked away from a successful run as a scientist in corporate America to teach high school.

The stakes of everything that happens in Andy's science research class seem to rise exponentially here in Greenwich, population 62,000, a beautiful seaside town that borders New York State and offers sweeping views of the Long Island Sound. It's a town beset with competition, even beyond the usual suburban jostling for bigger houses, better cars, and sterling children. But beneath the readily apparent affluence, Greenwich is a multifaceted place—and a lot more diverse than you might think. At Greenwich High School, 21 percent of the student body is Hispanic, 9 percent is Asian, and 4 percent is African American. Greenwich also has considerable populations of immigrants and expatriates; GHS is home to kids from sixty-one countries. Twenty percent speak a language other than English at home.

Andy's class fully embodies this diversity and spans the poles of Greenwich privilege. He's had a student whose family spent thousands of dollars to outfit a custom lab in their basement, as well as a kid to whom he offered a sweater, thinking he could use what for Andy was a holiday castoff. He regularly has students from families for whom money is so abundant, it's an afterthought. By contrast, he teaches kids who live so far outside the money bubble, their families rent modest condos because purchasing a home in this town—the median home price is $1 million—is nowhere within reach. One year, a kid asked for an extension to pay the initial $150 project fee because he was covering it from his after-school job earnings. And when it's become apparent that a kid's family can't pay for some of the extras, Andy finds a way to pick up the costs without ever mentioning it to the student.

The common thread among all these families—the multigenerational Greenwich establishment as well as the new immigrants who live (figuratively and literally) far from the country clubs, rowing clubs, polo grounds, and lush estates—is that they want their children to get a superior education and have chosen Greenwich for this reason. But in so doing, they have entered into a uniquely competitive culture of achievement—one that plays out in Andy's class.

The science fairs hold a powerful draw for Andy's students. For starters, there is prestige attached to even getting accepted to compete in a fair, where kids display sleek posters and give daunting oral presentations about their work. They stand in crowded, buzzing auditoriums, dressed up like mini adults (jackets and ties for boys, pumps and skirts or dresses for girls), fielding a barrage of questions from judges, as if defending a dissertation. The entire experience brute-forces them into a solo public performance where they expound on the intricacies of their projects, every word their own, propelled forward by the culmination of their hours in the lab, hard work, frustration, and, finally, show-worthy results.

The fair experience is arguably one of the best facsimiles for an adult high-pressure situation, made more so because many competitions ban parents and teachers from the actual arena. So there are no supportive glances, open arms, and big hugs—or meddling. This isn't like being on the debate team, which has the rah-rah aspect to it. Or being in a school play, where even if you fumble, someone else steps forward to fill the silence. No, these kids must demonstrate total mastery over their complex inventions and the underlying science. They have to describe in detail, say, the construction of a tiny polymer vessel loaded with cancer drugs and metallic particles that is guided directly through the bloodstream to a tumor by a handheld magnet over the skin. (That was an invention by one of Andy's star veteran students.) And the science fair format can be a harsh about-face from Greenwich child-rearing culture, where, as one teacher put it, "every parent thinks their child is a special flower." The judges don't fan the flames of ego or puff the kids up. They take critical aim at their work, to insure the competitors both know the content and were actually the ones who produced it.

In some of the entry-level fairs, the prizes are quaint: $25 Amazon gift cards or free museum admission. But as kids ascend through the fair circuit, they often vie for large sums of money—particularly in the über-elite contests where Andy's kids compete—with individual payouts as high as $250,000, or the equivalent of a free ride to college.

And beyond the money is the brief but bright light of fame they enjoy—going on late-night television, giving TED talks, getting invited to the Nobel Prize ceremonies in Stockholm. In all of these venues, they are cast as whiz kids, special beings whose intellects belie their ages.

Andy is constantly besieged with snickers and commentary about how nice it must be to teach in such a cushy town—the implication being that the affluence gilds the road to science fair victory. The eye roll at the mention of Greenwich is one thing, but the suggestion that Greenwich's wealth is the reason for his kids' success is infuriating to Andy. There is never a shortage of encounters with people who imagine he has carte blanche to buy whatever his kids need, that he is coated in monetary pixie dust, and that his competitive edge stems from a phalanx of kids with plump piggy banks.

The reality: for the science research program of roughly fifty kids, Greenwich High School's science department allots Andy a grand total of $1,200 per year.

The news about Shobhi and William was a welcome start to Andy's vacation. He had just begun to settle into the rhythm of summer, which for him always presented a crisis of sorts. Andy was many things, but idle wasn't one of them. The insane pace, hours, and agony that he devoted to his kids throughout the year kept him at his natural stasis. He thrived on the craziness. So the arrival of summer didn't see Andy sitting in a beach chair with a beer cozy and a magazine. Instead, he paced about like a freshly released animal that didn't know it was free to roam beyond the confines of its cage. The freedom felt like a death grip, especially that first week without kids and his manic amounts of tasks.

But by July 16, he'd reached a relative state of relaxation. He managed to log a beach day or two. He'd just splurged on a 2014 BMW F 800 GT motorcycle, which he'd been coveting for years. The bike was a fun and useful purchase because it provided a whole new dimension for one of Andy's favorite pastimes: tinkering. He had an entire side business devoted to it, where he repaired and calibrated the kind of scientific instruments he'd worked with during his corporate life.

Since he was never capable of total detachment during the summer break, he spent a sizable portion of his mental energy thinking about what lay ahead, going over his roster of students, making minor predictions about how their year in the class might take shape.

The demand for Andy's science research class was such that he made students apply for the roughly fifty spots. Every year he had to turn kids away, and over time, Andy had gotten better at the filtering process. He now puzzled over Romano Orlando, a junior who'd produced the single worst project proposal he'd ever received, an odd treatise about the need to make Greenwich High School green. It was Andy's custom to jot notes to himself on the pitches as he worked through the stack. Romano's was such a head-scratching combination of bad and weird, Andy just scribbled question marks at the top. And yet he decided to accept Romano, largely on the basis of a recommendation from the kid's Honors Chemistry teacher.

There was Danny Slate,* the charming, brilliant, and good-natured kid whose junior year project resulted in total collapse for a variety of reasons. Andy couldn't shake the feeling that he'd be utterly disappointed if Danny, whose senior year would mark his third with Andy, exited the class without having completed a substantive project. In the spring, following the implosion of Danny's attempt to link a biomarker called NAMPT to an increased risk of colon cancer, Andy and Danny read a scientific paper that suggested portobello mushrooms cooked at extremely high temperatures could generate battery

---

* At the request of his family, his name has been changed, as have those of his family members.

power. The project appealed to Danny, an amateur chef as well as a budding science mind. He liked the novel aspect of it: mushrooms plucked from the produce section and cooked into batteries.

William was coming back, and Andy was eager to see if his star student would pursue a brand-new project, something few seniors did. William's sister, Verna, would be in the class. Verna was a kitten of a kid, a shy sophomore with braces and smiling, bright eyes that revealed much more than Verna was willing to say out loud.

Also returning would be Olivia Hallisey, another of Andy's most winning students. Olivia had become somewhat of a teen celebrity after her big score at the 2015 Google Science Fair, where she won the Grand Prize, collecting $50,000 in scholarship money and a nifty trophy made out of LEGO blocks, a nod to one of the major competition sponsors, for a cheap and novel Ebola test she'd invented. Olivia was part of the Hallisey family science research dynasty that was launched by her two older brothers, Will and Blake. The youngest Hallisey, Charlotte, was a freshman and Andy had heard that she might petition to join research.

Sophia Chow, a shy, brainy girl who'd produced a deeply personal biology project her sophomore year, was coming back, and Andy had high hopes that she might be able to expand her work even further to make it really shine. She'd made it to only one fair the year before, for a host of reasons beyond her control, so he thought if her research panned out, she might be able to make some noise on the bigger circuit.

The 2016–17 class would be a mix of veterans and newcomers—those who were wise to what it took to make it in Andy Bramante's science research class, and those who had no real clue about the challenges that lay ahead.

2

William

During his junior year, William Yin had read about a trendy meal replacement drink that was popular among Silicon Valley entrepreneurs. Soylent was hawked as having all the vitamins and nutrients you need, providing a two-thousand-calorie-per-day diet. It got a reputation for being a huge time-saver for enterprising people developing products and technology so critical to mankind that they could cease eating and just drink their meals.

William quickly calculated he could save about seven hours a week leading the Soylent life and ordered a bulk shipment. He'd replace several meals a week with this liquid wonder.

"It first started with powders and then it was bottled drinks. People would know me as that Soylent kid," says William, adding a mild splash of a sales pitch with, "Five bottles a day, plus water to stay hydrated. You don't have to take time to eat." He emphasizes that last point as if saving the best feature for last.

"Oh, my god. The Soylent," says Andy, laughing at the memory. "We'd be standing around eating pizza and hold it in front of William, going, 'Come on, you sure you don't want to have some pizza?'"

William never broke. Like some kind of loyal cult member, he espoused the benefits of his liquid meals and his increased efficiency, and he even ventured to say they tasted good. He now says he worked fairly hard to convince himself of that last detail.

William sniffs. "Mr. B. isn't willing to sacrifice taste for convenience."

William certainly liked the mental endurance aspect of Soylent—and being the butt end of so many jokes just made him double down on his determination not to eat that pizza. He wanted to stand out for his invincibility. And part of that was feeling like he was made of the stuff of the Silicon Valley crew—that he, too, could scrap food, increase his efficiency, and kick ass.

It lasted several months, an admirable run.

The most precious commodity in William Yin's life is time. There simply isn't enough of it for everything he wants to accomplish.

Even in the pantheon of smart kids, William stands alone. By the time he graduates in June 2017, he will have taken twenty Advanced Placement exams, the college-level, subject-specific tests that can get kids college credit before they ever set foot on campus. He self-studied for a quarter of them. The school schedule is such that he wasn't able to actually take all twenty AP courses.

To put the twenty AP exams in context, consider this: an extremely driven kid at GHS might take ten AP courses throughout his or her four years. (The school strongly discourages freshmen from taking any of the nearly thirty AP classes it offers.) But there's no policy around self-study. No one's policing kids who choose to teach themselves a year's worth of college-level subject matter.

"I liked the ability to learn stuff on my own without the confines of, like, teacher-imposed deadlines, and homework, busywork, that would often come with an AP class," says William. "And freshman year, when I learned [AP] computer science, which was a decision that I made at the last moment, I really liked being able to sit at home and work myself through the computer science curriculum and look online, do my own practices, versus actually taking the course and following these guidelines constructed by the teacher, essentially."

He got the highest scores, 5s, on all but three exams. He felt a bit bruised by his 4s, despite the fact that they can still earn him college credit. Not being overly prideful, he chose to actually take the AP Spanish class after getting a 4 on the exam. He wanted that 5.

In GHS's eight-day, forty-eight-block cycle (a block being a fifty-eight-minute class), a student typically takes five or six classes—and has any given class six times in the eight-day cycle. As a sophomore, William went to see Suzanne Patti, his guidance counselor, with a list of the twelve classes he wanted to take.

She flat-out told him the answer was no. So he left her office and embarked on a fact-finding mission to try to fashion a custom itinerary that would allow him to take everything he wanted. He surveyed his friends and teachers to get a grasp of the school's complex master schedule. He tried to broker deals, asking teachers if he could attend some classes half-time and do the lab portion of chemistry apart from the rest of the class. The following week, he was back in Ms. Patti's office with a detailed proposal that allowed him to take all twelve classes. She reluctantly relented.

During his junior year, he took ten courses and got straight A's, as he had freshman year. His habits and results stunned even the rest of the brainiac set at Greenwich High. Soon William transcended this set and his reputation took on somewhat of a mythical halo. The kids were quick to point out that William was his own thing.

Danny Slate distills the whole Yin phenomenon as such: "If you try to compare yourself to Yin, you will lose every time."

William's drive has always been a source of chatter among the academic elite and their parents. People puzzle over what's motivating him, what makes Yin (as he's known and called) tick. The low-hanging assumption is that he must live in medieval-torture-chamber circumstances with his academics-obsessed Chinese immigrant parents standing over him, watching his every move until he produces pristine work. But William proclaims that the driving forces are totally internal. In fact, he feels grateful that his parents, Mark and Wendy, cut a wide berth around him. They tend to watch from the sidelines, supportive but not meddling.

"A lot of it is my drive to achieve as much as I physically and possibly can. Like, I want to hit the limit, and that's what I'm trying to do," he says. "I feel like it's fun to have this extreme level of stress. There's a certain thrill to it."

In addition to the adrenaline junkie aspect to pushing himself to berserk levels, he also admits that he derives pleasure from standing atop the achievement mountain at GHS.

"I've sort of established a reputation and I want to maintain it," he says. "There's this feeling that everyone sort of knows my name because I'm that crazy kid who took, like, twenty APs, or whatever."

Total domination, however, does not always lead to fulfillment. While William cut a swath through Greenwich's academic elite, he only truly felt intellectually ignited in Andy's science research class. Nearly everything else that school required of him felt like busywork, the mental equivalent of four years' worth of needlepoint. Taking Honors Chemistry his sophomore year was a waste of time, he told Ms. Patti. He should have skipped it and gone right to AP Chemistry.

What drew him to Andy's lab was the independence to take an idea and devote himself fully to the intoxicating and maddening process that is science.

"I felt like research offered me a freedom that I couldn't find anywhere else in the school," he says. "I don't find as much of an ability to think outside the box in most of the courses that I take."

There were no boxes that needed to be ticked, no tests, just time and space for his mind to dissect and attack scientific quandaries of the highest order. Yes, there was a scientific method and specific ways to execute complex processes, but William was a visionary, someone whose mind could see around corners in ways that often eluded adults.

Andy likes to point out that young minds are primed for scientific innovation. Kids don't know what they don't know and haven't had as much mud splattered on their faces when it comes to disappointment and failure. Their teenaged tunnel vision can be a gift when it comes to the risk-taking and boldness that best serves science research. They plow forward, undaunted and unvexed by life's pitfalls and rejections.

History is populated with examples of scientific and technological innovators who achieved great things in their youth, from Newton to Galileo to Einstein, right on up through to Gates and Jobs and Musk. Certainly with the contemporary examples, these iconoclastic men were propelled forward by outsized smarts, yes, but also a certain myopic belief in their ideas, as well as life experiences limited by the fact that they were so young (early twenties) when they launched their first inventions.

While Andy isn't expecting or necessarily shooting for the next Einstein to come out of his lab, he drafts off the tailwinds of his kids, who often embark on their research with a youthful enthusiasm and energy (and naïveté) that is contagious. It's a symbiotic relationship because when they sense he's game, they ratchet up their efforts. This had certainly been the case with William. And through the priceless relationship he'd developed with Andy, he had gotten so much more.

He had been accepted to the class as a sophomore, a time when many of Andy's kids aim to produce a good starter project. Inaugural projects tended to take on simpler problems, like removing secondhand cigarette smoke from indoor air. William, by contrast, explored a few things before turning his attention to cancer treatment. He developed a cluster of nanoparticles that he engineered to perform two key functions: enhancement of radiation therapy and efficient drug delivery. He handily won a top prize at the CSEF, which earned him a spot at Intel. There, his project won second prize. (Of the roughly

1,800 competitors at Intel, only a few hundred go on to actually place.)

Fueled by Andy's class and the electric experience that is Intel, William started thinking of a project for his junior year. The summer before eleventh grade, he watched a surgery involving a patient with severe plaque in his carotid arteries and was left transfixed by a simple question: Why was there no cheap, easy test to screen people for the frightening predicament of walking around with gummed-up arteries, zero symptoms, and a strong chance of dropping dead from a heart attack or stroke? The stealth nature of such a dangerous condition haunted William. There had to be a better way.

In his mind, he could envision a Band-Aid-sized sticker for the neck that would tell if someone had a dangerous level of plaque in his arteries. The model would involve nanoparticles, white blood cells, and pigskin as a stand-in for human skin. And he wanted it to be cheap—more affordable than an ultrasound or other expensive diagnostic technologies.

When William went to Andy with his project, Andy could not quite believe what he was hearing. The design showed outsized enterprise on William's part: He'd found a number of methods and practices and stitched them together to devise a truly innovative process and product. In theory, it was mind-blowing. Would it work? Andy had serious doubts.

The need for pigskin sent William on a Kafkaesque expedition, which involved many people thinking he was a crank caller and slamming the phone down, followed by a trek to a Manhattan butcher— the only place that agreed to sell him the stuff. He brought the frozen skin to Andy's lab, only to discover it was twenty times the thickness he needed. He made several attempts to manipulate it to his specifications, all of which failed. He and Andy stood over the roll of pigskin, spread out like a runner across the lab table, and puzzled over what to do.

They eventually decided to try sausage casings that could be bought at ShopRite, and that Andy thought might be the proper thickness. William had recently gotten his driver's license, so Andy sent him to the grocery store in the Yin family minivan. "Now, there's a chick magnet," Andy told William one day when he saw him tooling around in the old beater. One of the things Andy loved about William was that he could speak and joke plainly with him without any sugar on top. He could tell William when William was being a pill and pissing him off because, like family, neither of them was going anywhere. Andy was pleased to see his protégé triumphantly walking into the lab the following day clutching a bag of sausage casings. William shoved the useless pigskin in his lab table cabinet and set to work handling the delicate casings.

After a stressful few months that led up to a characteristically panicked two-week sprint to finish the project, William's invention took a top prize at the 2016 Connecticut Science and Engineering Fair, sending him to Intel for the second time, where he again won a second-place award. But at the prestigious global competition known as ISWEEEP, William not only won a Gold Award, he took home a Grand Award, which is given to the top invention in each of four different categories. His Grand Award earned him a trip to Stockholm to attend the Nobel Prize ceremony. Since he was only seventeen years old in December 2016, the time of the awards, he was able to defer the invitation to the 2017 event.

As proud and pleased as Andy was with William's success, working with William would come at a huge personal cost to Andy. William was so overextended, he showed up in Andy's lab only when he could fit it in, something that swiftly pulled Andy's irritation levers. (Andy is required to be in the lab when the kids are there.) William would stroll in at four P.M. or later, after he'd satisfied his other commitments, ready to clock a marathon session. Andy would lecture William about how the constant assumption that this was okay was an abuse of Andy's time and goodwill. He wanted William to be fully cognizant of every concession he made and that, on some days, those concessions were exclusively for William. Other, less encumbered kids were more organized and efficient, whereas William worked on William time.

He sometimes went incommunicado. But as deadlines approached, he would skip all his other classes and turn up in the lab early in the morning and stay until midnight. And then there were the hoarder tendencies, the pained look on William's face when Andy would try to toss things like bubble wrap. Even for a careful and frugal guy like Andy, William's inability to throw out a corner of tinfoil was extreme.

"No! I could use that," William would say, his arm reaching for junk headed to the trash. Kids took pictures of his scarily messy lab table and dubbed it "Yinsanity."

One day in the spring of 2016, Andy and the kids were nearly flattened by a creeping, foul odor in the lab. It was so vile it practically brought tears to their eyes. Andy walked around trying to identify the source of the stench. As he got closer to a lab table, he had a flashback. He flung open the bottom cabinet, and there, right where William had left it, was the bag of now decomposing pigskin.

Oh, my god, thought Andy, horrified, but laughing. *He'd be a genius if he wasn't such an idiot.*

One of the more enigmatic aspects to William is how his many achievements and self-declared near mania stand in such stark contrast to his outward manner. He speaks slowly and at a volume that barely rises above a whisper. Andy likes to joke that when all the other kids are shouting, "Mr. B.! Mr. B.!" William comes in with his throaty, whispery, ever-so-subtle, "Misterrr Beeee," which could be almost frightful if you didn't know him. He walks slowly too, lumbering along as if there's zero urgency to anything he's doing. (He once got separated in the Manhattan subway from Danny and a bunch of kids because he just doesn't move that quickly.) And there are other contradictions about William that emerge in the most unexpected scenarios.

Andy had been stunned the year before when he took William to Dunkin' Donuts and William was completely paralyzed in the ways of ordering fast food. He stood wide-eyed before the racks of donuts, not only unable to make a choice but overwhelmed by what the choices were. It was as if he were a Martian deployed from some far reach of the galaxy to learn the intricacies of ordering a donut.

"What . . . should . . . I . . . get?" William asked Andy.

"Uh, whatever you want," said Andy, tapping his foot. They were headed to the airport for a major competition.

"But there are so many choices. What do you think?"

"Well, you could get a donut. That would seem to be the popular choice here," said Andy, growing uncomfortable as the girl behind the counter stared at William, who was oblivious. "Or a bagel."

"But if I get a bagel, what would I get on it?" William said.

And so it went. How on God's green earth could a kid who fashioned custom nanoparticles not be able to conquer the act of ordering a baked good? It was these moments that drew Andy close to the eye of the teenaged tornado and gave him that extreme sense of warmth, bemusement, and fulfillment. In the end, William ordered a goopy mess of a hot everything bagel, dripping with cream cheese, that stank up the inside of Andy's pristine Volkswagen convertible. He loved so much about William and didn't want to embarrass him, so he said nothing. Asking a kid to master the Beer-Lambert law to understand light movement through particle matter was one thing. Asking him to use a napkin and take small bites could cause unspeakable shame. Andy felt mildly ill as he watched the poppy seeds and onions flick off the bagel, decorating his seats and floor mats with William's tortured selection. He was so distracted by the smelly dripping bagel, he missed the exit for the airport.

3

Danny

When you arrive at the castle on Brookside Drive, one of the first things you notice is that for all of its splendor, there's no front door.

Turning onto Brookside from the Post Road, you pass an eclectic bunch of homes, unremarkable colonials and restored farmhouses. But at an abrupt bend stand two white stone pillars topped with lanterns. You drive through them and slow your car to a crawl, as you now find yourself in a dense bramble, the stuff of sprites and bubbling cauldrons. The road brings you to the castle façade, the center of which is hollowed out by an archway leading to an interior courtyard. The minute you step outside of your car, you're disoriented by the immensity of the place, the massive stone towers flanking what might be the castle's corners, towers that stretch to the treetops and look like gargantuan chess pieces.

You resist the urge to shout "Hello? Anyone home?" because given the vast structure before you, even a foghorn would be futile. So you hesitantly tiptoe about, and soon your walk begins to feel like a prowl. Despite the forbidding stone walls, the stained-glass windows, and the lack of anything resembling the twenty-first century, you have to assume the place is wired and someone is laughing at grainy-image you lurking around in search of the front door, or any door. You think, *This is a castle. Where's the thick oak portal with iron hinges and a lion's head knocker?*

You finally make your way to a small, unremarkable brown door tucked into a stone wall. Your eyes scan the doorframe and you spot a decrepit button that would seem to be a doorbell. You press it and hear nothing. You're not sure whether to touch it again. If it does work, you don't want to be that person repeatedly pressing the button. You continue to skulk around, listening for signs of life—voices, music, kitchen clamor, anything.

With no response to the questionable bell, you walk through the arch and into the courtyard, which has an acoustic perfection. Against magnificent silence, you can hear every scratch of your footsteps, the lilting of your breath, the percussive shiver of the trees. Head tilted skyward, you do a slow-motion 360, like some rapturous tourist in a promotional video for castles. You're still trying to take it all in when you hear a door creak open, the windowpanes rattling.

You bellow, "I'm not trespassing! I just couldn't find the front door!" You're embarrassed, bumbling. You fear you've unintentionally landed the role of court jester; the joke is on you. But you're elated to have found a human. Finally.

You've arrived at Danny's house.

Science has been a constant in Danny Slate's life—something he was drawn to from a young age and for which he has a natural affinity. Some of this is because science was foisted upon him in a somewhat brutal manner while he was an infant. While his mother's extended family was spending the December holidays in Saint John, she looked down and observed that her nine-month-old baby's face had grown red and swollen to abnormal proportions. The family summoned the resort doctor to tend to Danny. They deduced he'd rubbed eggnog on his face.

Eventually, a diagnosis was made: egg, peanut, tree nut, and shell-fish allergies. Food allergies of this nature aren't the same as having, say, an intolerance to dairy that leaves someone with a stomachache after eating ice cream. Danny's allergies were much more serious, potentially fatal, a fact he had to learn to navigate as a toddler. He can't remember a time in his life when he didn't read food labels, carry an EpiPen, or have a full scientific understanding as to why something so innocuous as a peanut could kill him.

It's impossible to know if science would have come so easily to him had he not faced the burden of his food allergies. But he loved the probing nature of science, the constant stream of questions, the close examination and lack of easy answers. Not only did the material not tax him, it invigorated him. Little effort is required for Danny to understand advanced levels of cellular function, the behavior of bacteria, or energy creation.

Where he sometimes gets tripped up is that science requires a fairly consistent amount of work. And here lies a pesky two-headed monster for Danny: selective laziness and persistent procrastination.

When Danny arrived at Greenwich High, there had been no question that he'd apply to Andy's science research class. And in so many ways, Danny was a superb fit.

Andy immediately recognized the brilliance, the verbal facility, the ease with which Danny could grasp and master most any scientific concept. But also apparent was Danny's tendency to shuffle his feet when it came to real work, to schmooze in class and not quite get the ball over the line—all the while well-meaning, but working from embers and not fire in his belly. It was not unusual for Danny to shine laser focus in the lab on . . . the *New York Times* crossword puzzle.

During his sophomore year, Danny looked at whether blood cultures were a viable alternative to antibody tests to detect Lyme disease, a project inspired by his sister's bout with the nasty tick-borne illness that has plagued the Northeast for decades. The disease is especially pernicious because it often starts with nonspecific symptoms, like fatigue, flu-like illness, and joint pain. Lyme tests are fickle and can be unreliable.

It was largely a data analysis project, one he mostly did with a physician uncle of his who was able to gain access to blood specimens that are nearly impossible for a high school lab to obtain. Danny showed that blood cultures are more effective in accurately diagnosing Lyme—and he took his project to the Norwalk Science Fair, a kind of starter fair where kids hone their presentation and public speaking skills.

For his junior year, he wanted to compete at the Connecticut Science and Engineering Fair (CSEF). He procrastinated, and in the end, up against the deadline for submitting a research plan to the judges, the first step for entering CSEF, Danny struggled to come up with a project. He listed terribly, throwing the ship of ideas in his mind into a state of tumult. He started with something pedestrian and in his words "not that interesting or even scientific," which was swabbing Chromebooks, the cheap laptops given to every student at Greenwich High, to see if Purell was effective in ridding them of germs (and whether they grew drug-resistant bacteria). He then slid in the complete opposite direction and pitched ideas that university-level scientists tackle and have not cracked. He briefly turned his attention to the idea of a vaccine for dengue fever, the tropical mosquito-borne virus. It became a rabbit hole of impossibles, so he moved on.

His next idea: he wanted to see whether an enzyme called NAMPT is a biomarker for colon cancer—that is, whether its presence in the blood signals an increased risk of the disease. This was a relatively novel and good question. The problem: it required data from human blood samples of people who'd had colon cancer. He made inquiries at medical institutions and not a single one would give him the data. His father even put in a request at a high level with a major New York hospital, but to no avail. Once it became clear that Danny's mission was for naught, his efforts wilted. With no shot at making any of the fairs, he retreated from the front lines and Dannied about.

Ultimately, the crowning achievement of his junior year was becoming the Pizza Man. What started as a little treat for him and his friends—having his family's nanny deliver pizza to some undisclosed location so Danny could smuggle it into Andy's lab—became something he did for the entire class once every eight-day cycle. Danny is both generous and has natural social grace—he knew that the right thing to do was to provide pizza for the masses. And provide he did. He never asked the science research kids for a dime; he made no show of this gesture. What he did relish was the illicit manner in which he managed to get the stack of steaming pizza boxes past the security guards—who forbade food of any kind from entering the school's science wing.

This did not go unappreciated by Andy. He saw that Danny was giving where other teenagers might be greedy or clueless. Plus, Andy enjoyed watching his students descend on the pizza like seagulls on a shoreline carcass, as if starved and unlikely ever to eat again. Every time, the pizza vanished in seconds.

Andy kept a watchful eye on all of his students—surreptitiously and overtly monitoring the various goings-on in their lives. He kept a mental log of their overall arc in the class. Danny's Lyme project was good but a complete outlier for both of them because it was largely done off-site. The cancer project died a slow, predictable death. And here was Danny, headed into senior year without anything that exemplified his capabilities.

So in the spring of junior year, Andy and Danny agreed Danny would pursue the portobello mushrooms as batteries project, where he'd cook the fungi at extremely high temperatures onto silicon wafers. If successful, they'd act as anodes—the positive charge for a battery. Danny liked the mushroom battery, even thinking he might try using a vegetable with more carbon, like carrots. That was Danny, immediately thinking about ways to expand upon a basic idea.

Though seniors rarely embark on new endeavors, due to college applications and senioritis, Danny and Andy decided that if Danny could develop the mushroom battery, he'd be able to apply to the Regeneron Student Talent Search, the big-ticket competition for seniors only. But much more important than the contest or the money, Andy wanted Danny to leave the class with a true accomplishment under his belt. Andy couldn't really bear to let any kid slip by, even if it meant sticking a fishhook in his mouth and yanking him up from the murky pond of inertia.

So in the spring of 2016, Danny got to work. Standing with a pile of portobello mushrooms, he shaved off the skin and baked them in the lab's high-temperature oven onto the two-inch-wide silicon wafers. During the heating process, the mushroom skins cooked and became completely flush with the wafer, creating silver pools on the surface, so-called nanoribbons, which are impossible to see with the naked eye. Their presence can be confirmed only by viewing the wafers beneath an electron microscope, a costly instrument that Andy doesn't have in his lab.

At the end of the school year, pleased with his wafers, Danny encased them in a little plastic container. He'd resume his project in the fall when he returned to the class.

4

Sophia

Sophia Chow was in seventh grade when her body started behaving peculiarly. Something felt very wrong, as if there was a kind of cruelty being unleashed in her system. She became hypoglycemic, making her dizzy and weak. Pain flared in her left hip, an intense soreness more common in someone three times her age. And then there was the flattening fatigue, where she would walk through the door from school, collapse onto the couch, and fall into a comatose slumber for hours before barely being able to lift herself to eat and do her homework. She quit lacrosse because she lacked the strength needed to run for hours on the field. She retreated more and more, unable to do much except be still and conserve her energy. She often took a pass when friends invited her to go out.

Then intense pain besieged her head. It invaded at unpredictable moments and was remarkable for its force and reach. From the top of her skull to the tip of her chin, Sophia felt as if her head were encased in a sinister grip. The insides pushed back, pressing against her skull, as if trying to fight their way out. Typical over-the-counter meds did nothing to ease this pain because it was anything but typical. And then she'd be tormented by bouts of insomnia, unable to sleep as she watched the numbers on her clock flick from two to three to four A.M.

Her mother, Stephanie, took her to the doctor multiple times. Alone, the symptoms were disparate pieces of an odd puzzle. Together, they didn't signal an obvious ailment. Whatever was dogging Sophia's system seemed to lurk around her body and hide for periods, only to reemerge with a vengeance. She'd have stretches of feeling more like her old self, but something just wasn't right.

Looking at Sophia, it's hard to imagine internal physical suffering in someone so head-swiveling gorgeous. As a biracial Chinese-Caucasian girl, she received striking traits from both parents, with high cheekbones and feline eyes encircled by lashes so long and luscious, they probably have dreams of their own. Beneath her beauty is a quiet elegance. She lacks any teenage flash or ornamentation meant to accentuate the fact that she hit the DNA jackpot for aesthetics.

She had always been a quiet, watchful, self-contained child with a mind organized like a card catalog. She loved going to Michaels and stocking up on arts and crafts kits. Sophia would sit by herself for hours, stringing and painting and gluing. When Gregg and Stephanie would take their young twins (Sophia's brother, Brandon, was born minutes after her) on vacation, Stephanie says that inevitably the hotel room would devolve to tornado-like conditions and she'd be unable to find anything. No matter how small an object, even as a four-year-old, Sophia could pinpoint its whereabouts. To this day, she's still the de facto family archivist, finding things for Gregg when he needs them and has no idea where they've gone off to.

She's always been drawn to pets, and given her serene demeanor it's easy to imagine her getting a lion to pour her a cup of tea. But despite her outer calm, internally there burned an intense desire to excel—and a constant fear of falling short of anything but academic perfection. She approached her schoolwork with such gravity that

Stephanie was constantly urging her to ease up and relax, the opposite of a tiger mom. (Brandon, a talented baseball player, was more chill in his approach to academics.)

For her part, Sophia fully admits she doesn't know how to be anything but a perfectionist. "I'm not so much competitive with other people but with myself," she says. What she'd learned was that she had a natural drive, if not a passion for any one thing. The competitive culture of GHS could do that to you. You became so embroiled in trying to achieve, you didn't even stop to consider what made you truly happy.

Her illness wore on throughout 2012, rearing its head without warning and then dissipating, but never fully disappearing. One day, as she and her mother were leaving the doctor's office, and the physician was saying she'd run yet another round of tests, Sophia stopped Stephanie and turned to her. Sophia had been considering every possible thing that could be sapping her—talking to people and gathering ideas. She said just one word.

"Lyme."

They locked eyes, pivoted, and walked back into the doctor's office. They asked her doctor to add Lyme disease to the blood panel.

The test came back positive. And so it was that the Chow family learned that it was Lyme disease that had been playing a cruel cat-and-mouse game in Sophia's system for the past few years. It's a nasty and often debilitating disease carried by the deer tick, which, because of its small size—about the tip of a ballpoint pen—is extremely difficult to detect. Not only was Sophia not aware she'd been bitten, she never had the hallmark sign that often, but not always, appears: a bull's-eye rash, a distinct red circle around the bite.

Following an initial sense of relief that they had uncovered the cause of the odd and painful symptoms, the family was terrified. They knew people who had been ravaged by the disease, unable to walk, their faces partially paralyzed. In addition to the fact that the severity of Lyme ranges greatly, the disease and how to treat it are controversial in the medical community. Some doctors don't believe chronic Lyme disease—a version that can wax and wane for years following a single tick bite—exists. But since so many patients continue to suffer symptoms after the standard two to four weeks on antibiotics, some scientists believe there's a condition called post-treatment Lyme disease syndrome, the symptoms continuing despite the absence of the disease in the bloodstream.

The Chows decided to take Sophia to a Lyme specialist in Wilton, Connecticut. He immediately put her on a six-week regimen of strong antibiotics, which Sophia had an extremely tough time tolerating. They left her nauseous, and after years of all her Lyme symptoms, she sort of crumbled in the face of a new set of painful gastrointestinal ills.

Then, one day during her freshman year at GHS, as she was still trying to get comfortable with the treatment, Sophia's biology teacher, Dr. Gambino, pulled her aside.

"I thought I was in trouble," says Sophia.

She wasn't. Dr. Gambino had recommended Sophia for Andy's class, which she'd never heard of. Her initial response: "That's so weird. I'm not going to do that." The free-form nature of the whole thing sounded daunting, so removed from the regular high school cycle of read, memorize, test. But after giving it some thought and talking to students in the class, she saw an opportunity.

She'd try to cure Lyme disease.

5

Romano

When Romano Orlando started playing football as a freshman at Greenwich High, he almost immediately got a bad vibe from one of the coaches, a guy who was widely known to be a hard-ass who mentally brutalized the players. People talked about him in terms of whether or not someone had what it took to handle him. If you could, cool. If not, you were a pussy, which the coach would openly scream at you.

Football had been a constant in Romano's life. He started out playing in the town's pee-wee league and kept with it right into high school. At five foot nine and a half and 155 pounds, he has a solid medium build, not particularly broad but all muscle. He played strong safety freshman year and shifted to free safety as a sophomore, on the JV team both seasons. By his own assessment, Romano wasn't an amazing player, but he was respectable and had been named MVP his final year in the town's club league.

The coach played favorites and was worshipped by some. But Romano sensed a targeted animosity from the guy, an ever-present seething, often silent, but other times very vocal. Not surprisingly,

Romano didn't love it when the coach barked "Pussy!" at him. Romano started to notice he had a complex cocktail of emotions around this guy and football. He carried an intense sense of dread, a desire to please, and a fear of falling short of the coach's expectations. On the one hand, he knew the dude was extreme, a near parody of the nut-job coach who somewhere lost sight of the fact that this was high school ball—these players, many of whom hadn't even started shaving yet, weren't being groomed for the NFL. Still, Romano couldn't quite shake the desire to perform and find good footing with this guy. It drove all his choices on the field, and not for the better.

"So the last game we played, the quarterback threw a pass to my section [of the field] and the kid caught it and, you know, we were up by, like, three touchdowns, the last play of the game," says Romano. "I could have just tackled him and, you know, be done with it, but I was so afraid that this kid made a catch in my zone, I was so afraid of what this man would say, that I panicked and instead of going in for the normal tackle, I went in for the big hit. But a lot of times that doesn't work and it didn't work. And he ran and got thirty more yards.

"They blew the whistle, the game was over, and I walked back to the sideline, my head was sort of down. Everyone saw it. Like, I looked like a dumb-ass, going for the big hit. Everyone kind of thought I was doing it to be an ass, you know, to hurt the kid, but I was just so afraid to mess up.

"Everyone was laughing. I looked up and I looked at the coach and hand to God, he stared right at me. For, like, two seconds, we locked eyes, and then he just looked back and kept walking.

"It was the look, the sort of Greenwich look-down-on-you look, it was the look where it was like in my soul, the look where . . . it was bad and once that moment hit me, I was like, okay, this isn't healthy."

Romano had considered walking away during the season, but Orlandos weren't quitters. Romano's parents, Kim and Rome, had instilled in their three kids the idea that quitting wasn't available to them just because something didn't come quickly or effortlessly. They didn't puff their kids up the way many parents did in this town.

No matter what, their kids needed to finish out a season and decide from there. So Romano had stuck it out, feeling tormented and never getting to that agreeable place with the coach. He'd been marked as a pussy from day one and that's how he was going to finish.

The thing was, all this bad blood, verbal abuse, head-game shit was so anathema to Romano. He was handsome, popular, and a walking study in cool. Not cool in that he was going to peak in high school kind of cool, but cool in a way that stirred you to think about the meaning of cool. It was like the famous Supreme Court writing on the definition of pornography: you know it when you see it.

Romano went against virtually every high school social instinct—ironically, the very instincts that drive the desire to be cool and protect social status. During the first few weeks of school, he noticed a kid with autism whom he knew from middle school eating alone in the cafeteria and invited him to sit with the popular kids. None of them welcomed the kid with open arms, but they also didn't dare say anything to Romano.

Romano could give a detailed breakdown of the GHS social hierarchy. In keeping with high school convention since the beginning of time, cafeteria geography reflected status. Every group had its place and there was nothing in the way of intermingling. Romano sat with the popular kids: the Neanderthal jocks and the blond queen bees. He caused a seismic rupture of decorum when, in tenth grade, he started dating a girl we'll call Caitlin, from a midpopular table. But he liked Caitlin, admired her moxie in pursuing him. Plus, Caitlin had the approval of Romano's older sister, Sophia, who vigilantly guarded her younger brother and was quick to disapprove of those wanting to date him, a line that was long and eager.

In the middle of his sophomore year, when Romano heard about Andy's class, he had a visceral reaction to it. "I was like, this class sucks," he recalls. "Six blocks, full year, screw that. Do you think I'm research-ing? That's some BS. That's some foofoo shit. I'm not doing that."

But then his take slowly shifted. "Once I started talking to my friend Haley [Stober] about her project, I was like, wait, you literally just research stuff? She's like, yeah. I was like, it must be so intense and she was like, no, it's like a relaxed environment and I was like, what do you mean?"

He went out to lunch with Sophia Chow, another friend, whose endorsement of the class inched him forward. But the turning point came when Romano attended the science awards. He was there be-cause he was on the science team (and also on the math team, two activities he wasn't eager for his friends to know about) and he watched with awe as the science research kids dominated the awards. Rather than viewing them as a bunch of nerds, he saw astoundingly bright kids whose punch came from their intellect.

"I remember that solidified it. Once I saw those kids get up there, I was like, those are the kids I want to be spending my time with. You know? Especially William Yin. He was on that stage more than Dr. Winters. I was like, that's amazing. I'm with all these jarhead kids playing football and this coach who was making me feel like an ass. I've never had a man be so mean to me in my life and then take Mr. Bramante, who was in the running for a Presidential Award. Wasn't he like top three for teachers? He was a finalist. So that's one-zero sci-ence research. And then I look at the kids and I'm like, they're so smart and they're so successful and they're nice."

So, with no pointers of any kind, he wrote up a proposal to try to get into the class. He was completely, obliviously out of his element—and it showed. His pitch to make GHS green was something a fifth grader might write. Andy read it and did one of those rapid cartoon headshakes. He read it again. It was the oddest and worst proposal of all time. It was sitting in a stack of pitches that included things like a cheap test for Chagas, a bug-borne disease endemic to poor areas of South America, and the creation of a saltwater battery that runs off a submerged turbine attached to a buoy.

Andy approached Cindy Vartuli, a sassy chemistry teacher known for her hard tests and even harder-won affection. She liked to tell people that a Greenwich mom once called her "the destroyer of young boys' dreams." Cindy grew up in Greenwich and graduated from GHS. But she was finishing her Ph.D. and was equally sharp in reading people. She didn't view Greenwich culture through any kind of soft-lit filter.

"Take Romano," she told Andy. She loved Romano. Yes, she and Romano bonded over being Italian American and they could both toss around the same ten Italian slang words. But he was fun, quick-witted, and a grade-A good egg. However, what really gave him premium standing was that Romano was an ace when it came to Honors Chemistry. Despite the fact that he didn't hang with the smart kid crew, Romano was a whiz in math and science, with effortless straight A's. However, he had absolutely nothing in the way of independent research experience. But Cindy loved him, and knowing that she wasn't easy to win over, Andy felt inclined to give Romano a chance.

Romano paid Andy a visit one day after school. Andy had never been dealt such a stinky toot of a proposal, and he wanted to somehow bring this into perspective for Romano. With his amiable exterior, Andy could pull off saying many things that would sound eviscerating coming from just about any other mouth. But from him, the comments actually became a much-needed icebreaker.

"Dude, what the hell is this?" he asked Romano, as if they were both in on a joke gone horribly wrong. "I have no idea what this is. What are you trying to do?"

"I know, I know," Romano told him. He confessed to whiffing because he truly had no idea where to get started. He'd later give the following assessment of his proposal: "It was butt cheeks."

But it was impossible not to like this kid. He could rival a power plant for energy output. He lived up to Cindy's hype. And it went far from unnoticed by Andy that Romano came from an entirely different social sphere than the usual suspects in his class. Romano might add an interesting interpersonal component to the gang.

So Andy filled Romano in on how to think about a project he might pursue. He noted that the plan should be scientific and not about school cleanup. He gave Romano marching orders to read certain websites with scientific papers to familiarize himself with the level of science needed for the class.

Romano was super psyched that Andy was going to take him, despite his butt cheeks proposal. He started thinking about what his junior year would look like. He knew that science research was a heavy time commitment and he wanted to do right by it. Slowly, some ideas started to emerge.

And by the end of summer, he had some other business he needed to take care of. He went to see the head football coach to get back the expensive helmet his family had paid for—and officially quit the team.

"He was like, this is going to be the biggest mistake of your life. You're going to regret this so much, literally. I'm like, no, I'm fucking not. I'm not regretting shit."

FALL

6

Andy

It's September 1, 2016, the first day of school in Greenwich. The Northeast has flipped its fall switch and the nineties temperatures of the previous week have fallen to the high seventies. It's a dreary, gray day with rain streaking across the windshields of the fleet of German cars in the student parking lot. The atmosphere bears none of the meteorologic promise you'd want for the first day of school.

Andy's class, room 932 of the school's science wing, is pristine, as it will stay for the first couple of months before the lab spread invades—projects living on every available surface. It's a large room, with six black soapstone-top tables, three high ones with stools that form a half moon in the front of the class, and three standard ones toward the back. The lab is populated with a bunch of topflight scientific instruments, such as gas chromatographs; UV, infrared, and luminescence spectrometers; HPLCs; and particle sizers—instruments that are almost never found in a high school classroom. Andy is a master at trading favors, and because he's got more goodwill in his karma bank than most, his friends and contacts in industry have helped him stockpile these used instruments that have aged out of the pro leagues but are precious to Andy's bullpen. Across the hall, he has a small, windowless room filled with more used instruments where kids often work or retreat to figure things out away from the din of the main lab.

He's in a sedate mood, not having yet mentally checked in. He and his family—his wife, Tommasina, and college-age daughter, Sofia—took a nice trip to Quebec ("like Europe, but you can drive there") with friends and the change of scenery had been good for him. But now he's back in his lab where he'll put in sixty-hour, six-day weeks once the science fair season kicks in. And while he loves and needs what he does, he's easing and not running into it all.

Sanju Sathish, a senior, comes bounding into the class wearing a backpack and bearing a box with a string handle. Sanju is stick-figure skinny with large dinner-plate eyes and a broad smile.

"Hi, Mr. B.!" he says, nearly out of breath. "I went to India this summer and I went to one of these markets that doesn't sell mass-produced gifts." He plunks the box on the lab table.

"Should I open it?" Andy says, eyeing the mysterious box.

Sanju tells him to go ahead and Andy unfolds the top and removes several layers of packing paper to reveal a beautiful eight-inch gold elephant statue mounted on a pedestal. The bejeweled and majestic beast is raising its trunk.

"Thank you very much," says Andy. "Wow. This is cool." He's both stunned and mystified, never having received an elephant statue before.

"I was in this bazaar and I saw it and I thought of you," says Sanju.

The previous year, Sanju had taken a protein in bee venom called melittin and showed that it can denature certain proteins found in HIV, which could have implications for treatment of the disease. His project was motivated in part by his annual summer trips to India, where he and his grandmother would often hike to a mountainous holy site in Tirunelveli, in the southern part of the country. Sanju was haunted by the emaciated beggars at the bottom of the mountain. Many of them had HIV, his grandmother explained, and no access to healthcare.

So the project had a poignant start, followed by a dead zone middle and then a backbreaking finish. Sanju lollygagged and waited until the last minute to crank up his engine. With two weeks to go until the fair, Andy cordoned off the lab and devoted an entire day to Sanju over February break, when many other kids were skiing the Rockies or sunning themselves in the Caribbean. Sanju didn't require heavy lifting from an intellectual standpoint—he needed, as Andy was fond of saying, "a cattle prodder in the tushy."

The last-minute push paid off for both of them. Sanju's project won second place in the health sciences category at the state fair, which earned him a ticket to Intel. When he learned he won, he texted Andy before even telling his mother.

The day Andy devoted to Sanju was burned into the kid's memory. He said it all came down to "the one day Mr. B. devoted just to me." He felt the weight of what he believed was a sacrifice on Andy's part and was awed that a teacher would do that for him.

"It was, by far, the nicest thing any teacher has ever done for me," says Sanju. "When I first came here, sophomore year, I was completely unconfident, I didn't think I had a place in the school. This class becomes your home, and Mr. B. becomes a father figure to some people."

So Sanju purchased just one gift while he was in India this summer, and that was for Andy.

This year, Andy's forty-eight research kids are split up into three sections—all with a very distinct vibe. For the second year in a row, he's also teaching chemistry to the English as a second language kids. One of his colleagues was desperate to rid herself of the gig and Andy's department head tapped him. The ESL set gave him a whole new perspective. They are nearly all recent immigrants and most come from families of modest means who have landed in Greenwich from South America, Europe, and Asia. Many are the kids of Greenwich's working class: the landscapers, cooks, and nannies. (There's also a sizable expat population in Greenwich, people in corporate jobs who have relocated here temporarily.) These students are the opposite of spoiled, and the families are often incredibly grateful to Andy for his work and devotion. Andy feels a certain connection to these kids, who he says remind him of himself when he was their age. Some days, the ESL class feels like a form of oxygen, a place where he can breathe his easiest.

Two of his ESL kids from the previous year cut homeroom to stop by the lab to say hi to Andy, as if they'd been away on a long journey. It so happens that these kids had been the ones who'd given Andy the biggest headaches. As they walk out, Andy shakes his head.

"The kids who are the biggest pains in the asses wind up being your best friend," he says.

His first class of the day is ESL chemistry, and it's a large, jolly group of kids, almost exclusively Asian and Hispanic. From a curriculum perspective, Andy can teach basic chemistry with little effort, and he enjoys it. Today, he uses two balloons, one inflated with helium and one with oxygen, to demonstrate basic principles of the weights of certain gases.

Homeroom at GHS is reserved for occasional days, like the first few days of school so kids can get some grounding in the 450,000-square-foot building, which includes a pool, a theater, and this year a brand-new $29 million music wing that houses an electronic music composition lab, a state-of-the-art recording studio, and a performance hall with pin-drop acoustics. There's somewhat of a music education mafia in Greenwich. Parents petitioned for the snazzy new music wing, and here at GHS, their kids can be in every ensemble configuration known to mankind. Which is admirable, but also more than a little ironic in a town where parents would rather eat carbs or drive a Chevy than allow their children to pursue a career in music. The mere suggestion of majoring in music in college yields a swift facial Morse code that is easy to decipher: professional music career + significant risk + subjective medium = likely life of poverty = no fucking way = we didn't move to Greenwich so you could major in music. This despite thousands of hours and dollars on private lessons and competitions to supplement the robust music program at GHS.

Still, the new music wing is a source of awe. Even Andy walks down the hall, peering at the shiny new practice spaces, veering off to point out the fancy auditorium. Studentwise, there's a lot of music-science cross-pollination, which poses major time challenges for students who do both.

Classes are slightly truncated this week, so Andy's first class, the ESL class, whizzes by. But this is an "A" day, which means he teaches four classes in a row, a grueling run.

His first research class strolls in and it's a mix of veterans, including Olivia Hallisey and Margaret Cirino, who both started with Andy as freshmen, and new kids, like Noni Lopez and Amit Ramachandran.

Kids scatter themselves around the tables. Andy stands at the front of the room doing a long, drawn-out roll call, taking the time to personally greet or catch up with each student.

"Christo! How are you, my friend?" asks Andy. "How were the college tours?"

"All colleges look the same at a point," deadpans Christo Popham, who's sporting a full beard and has a mad scientist–survivalist look going.

Andy asks Nick Woo, a tall Korean-American sophomore, "What's it like living with Derek?" Derek is one of Nick's older brothers. His project last year looking at the mysterious disappearance of honeybees had a surprisingly lackluster run on the science fair circuit, despite tackling one of the most pressing ecological quandaries of our time. Derek is known for being brainy in nearly all things school related but flighty when it comes to personal matters. (He famously texted a girl to invite her to the prom with one misspelled word: "Prum?") Andy always says Derek is at great risk of tying his shoes together. One of the more shocking and distressing things to Andy, a lifelong car aficionado, is that Charles Woo, patriarch of the Woo clan, is a retired hedge fund manager turned car collector with a notable stash of Porsches—and his sons have zero interest in cars. Andy practically begs them to take an interest.

"My god, if that were me . . . if I were your age and my dad had a garage full of Porsches . . . ," he says to Nick today.

After the catchup session, Andy gives a quick speech about the class, describing the self-directed nature of the operation and that due to the fact that he's a one-man band tending to forty-eight kids, there's a squeaky wheel element. Those who put in the time and effort are most likely to get his attention, which remains very divided.

He quickly acknowledges that the class has garnered a bunch of high-profile awards, and then he starts going through the various competitions and deadlines.

For the veterans who have completed, show-ready projects from last year there's a deadline of September 20 for the Siemens Competition.

"For whatever reason, and I don't know what it is, we've never really gotten traction with this competition," he tells the kids. "It's not like the other competitions, where we do pretty well. We haven't cracked the code here yet. So bear that in mind if you decide to apply."

At 11:33 A.M., as the class is near finished, a petite Asian kid with an elfin face walks in with a look of bewilderment, as if she's standing in the center of a bustling European train station. She's wearing giant white platform shoes that look like small yachts on her feet.

"I don't know what block it is anymore," she announces.

"Madeleine, really? Come on. You're a senior," Andy says to Madeleine Zhou, a science research legacy. Years ago, Andy taught her brother, Felix, who's now pursuing medicine.

He returns to his introductory speech, finishing with a motivational thought that doubles as a warning: "You only have one shot with your story at the state fair."

The next block of the day rolls in and it's a killer. Twenty-two high-octane kids. In walk William, Shobhi, Derek, Sanju, and plenty of newbies, too, including Romano, Verna, Peter Scott, Emily Philip-pides, and Collin Marino.

Andy does his meet and greet and delves into his stump speech, trying to set the stage for the year.

"The reality of this class is it's much more immediate. It ebbs and flows. At other schools, kids in science research get sent out to labs at universities or in industry. Here, you're with me, like it or not," he says. He gives the kids his cellphone number and tells them not to abuse it—which they inevitably do, texting him late into the night for all matters serious and trivial.

He walks a careful line with the competitions. While he needs to lay down the stakes to help get the inspirational juices flowing, he doesn't want to place too much emphasis on winning. He knows that's why many of them are here, drawn to the prospect of victory and money, but there's no benefit to ginning up that side of things. Andy's learned that if the kids are blinded by their pursuit of acco-lades, they get badly burned when the outcomes aren't what they hoped for. And then the whole thing feels like a heap of misguided intentions—like a road trip gone very wrong.

"You're going to learn a lot in this class that's science, but you're going to learn a lot of life lessons, as well. You will get out of this class what you put into it. You need to know that there are only so many of you that can go through that funnel for the fairs. Thirty of you will apply. Fifteen to seventeen will get to go to the state fair."

Andy's not a lecturer, so he keeps his remarks to a minimum and then asks the kids to go around and introduce themselves. He starts with William Chen, a senior who he accepted to the class, something he rarely does. Though he's new, William sits with his fellow seniors.

"I play the violin, I'm one of the science team captains. All of you should come join the science team," says William.

"Cheap plug," says Andy. "Now, the newbie table."

A brunette with a small voice says, "I'm Emily. This is my first year in research. I'm from a small country called Cyprus. Most people don't know it, so I just say Greece. I used to play violin."

Verna's next. This is also her first year of research and she mentions she plays the piano and runs. Everyone knows she's a Yin, which somehow makes people feel that this is really all they need to know.

When it's Sanju's turn, he says, "I'd like to do a fun fact. I was born on the ninth day of the ninth month of 1999." Fun indeed. But he's opened himself up with this public display of dorkdom. The kids pause and then immediately move to snuff out his birthday glory by pointing out that he's not the only one in their class year with this distinction. There's at least one other 9/9/99-er.

The bell rings and it's not an actual bell, more like a department store tone prior to an announcement. The kids amble out of class and Andy braces himself for the final research crop.

In walks Danny with his cronies, including seniors Henry Dowling and Takema Kajita. During the introductions, Danny says, "I'm Danny, also known as El Capitan or the Pizza Guy. I'm a senior and I'm captain of the math and chess teams. I'm also in the Latin club." Math team, chess team, and Latin club—a trifecta of things that in another era would get you stuffed into a locker. But Danny is quite proud of his extracurriculars and lists them head held high, his nerd charm on full display.

After the quick bio sketches, Andy says this about the class, "I like to say I'm training my future colleagues. This is about learning how to think independently. It's not about the science fairs."

But he wants to make one thing clear. "I get nosy. I ask you about things you probably don't want to tell me, like the prom."

Because from day one, when Andy's not thinking about the effects of biochar on plant soil or using fluorescence to chart cellular change, he's thinking about his side gig as a prom date broker. He's legendary for getting dates for anyone who wants to go to the prom, even the most un-prom-inclined kids. What better time to mention it than the first day of school?

7

Danny

September 21 is an eye-squintingly bright day, with remnants of summer humidity thickening the air. Danny and Andy are standing on the corner of Ninety-ninth Street and Madison Avenue, amidst a sea of scrubs and white coats outside Mount Sinai Hospital. Danny downs a cup of Dunkin' Donuts coffee before they head inside.

Early on in his gig at Greenwich High, Andy received a fortuitous offer. Dr. Ming-Ming Zhou, a research scientist at Sinai, told Andy to let him know if there was ever some equipment or technology he needed for the class but didn't have. Dr. Zhou's son, Felix, was in Andy's class and now his youngest child, Madeleine, is in her final year of research. Andy knew that the use of a scanning electron microscope (SEM) would take his kids' work to another level. An SEM captures three-dimensional images magnified by as much as one hundred thousand times, meaning it can make a dust particle appear like something that will decimate the human race. Dr. Zhou connected Andy to Dr. Ron Gordon, Ph.D., who runs an SEM at Sinai. Ron lets Andy and his kids use the instrument to view samples and capture images of their work—free of charge and mostly whenever they want to come.

Danny had cooked his portobello mushrooms onto silicon wafers last spring and now needs to see if the nanoribbons have formed. The research paper he's using as a guide indicates that the formation of nanoribbons is a critical step to eventually producing battery power. And the only way to see this is through an SEM.

The pilgrimage to Sinai is somewhat of a rite of passage for Andy's kids. They go with him solo or in a group of no more than three or four, which means much more Andy to themselves than they'd otherwise get. There's a palpable shift in dynamic when he and his kids get out of Greenwich. Everyone seems to take a deep breath and loosen up. For one, the hour-long trip, either by train or car, means time to chat about things other than science, to gossip, joke, and divulge. There's almost always some kind of food involved, a side jaunt to a restaurant someone has heard is great. So it's like a personalized platinum field trip, but with the slightly devious feel of playing hooky.

As fond of Danny as Andy is—because not liking Danny is somewhat impossible, akin to not liking baby koalas—in two years of having him in class, he hasn't had a moment where close connective tissue has formed. He doesn't know a ton about him, and since all the other kids are at the very start of their projects and months away from needing the SEM, he's glad to have this day alone with Danny.

Danny absolutely thrives off the city. He loves his ever-expanding know-how and comfort, his growing ability to blend and betray no traces of suburbia. He spent the summer in a research lab at Rockefeller University and loved walking back to his grandparents' apartment late at night, as if returning to his natural habitat. The city's main attractions—energy, anonymity, edge—are all a draw for Danny. They go a long way in eradicating the inertia that can frequently embalm him in Greenwich. So today, he's got a spark in his eye as they approach the lab. Plus, he has a specific goal: he'd like to complete his mushroom battery so he can apply to Regeneron STS in November, the seniors-only competition that requires, among other things, a finished project.

Andy stops by Ron's cramped office, introduces Danny, and makes some jovial chitchat. Ron's got a classic New York mensch personality. He's so far from pretentious, you wouldn't be surprised to hear he repairs mufflers at Jiffy Lube. He likes to tell people that his daughter and now son-in-law were on *The Amazing Race,* a natural icebreaker for Danny, who became a huge fan of the show while a toddler. Ron loves what Andy does with his students and is happy to lend his microscope to advance the cause.

Andy and Danny head to the small cutout of a room that houses the SEM. At a minimum, it's a $100,000 instrument, so Andy always does a demo for the kids before he lets them take the chair and start maneuvering it.

Andy fires up the scope and shows Danny how to load a sample for viewing. The microscope itself is in a vacuum so that electrons can be shot through the sample at high speeds. Specimens are mounted onto little stainless steel pegs and placed into a chamber for the high dose of electrons. Images appear on a regular computer monitor and there are several knobs and dials that allow you to zoom in and work around the sample.

Danny's silicon wafers are thin, brittle disks the size of half dollars with visible, dried chemical pools on the surface, resembling an antifreeze spill in water.

"Danny, do you watch much TV?" asks Andy.

"Is the pope Catholic?"

Danny talks about TV in a way that suggests it's one of his vital organs—and any attempt at one-upping him on shows watched or hours logged is a losing game.

"I remember watching the pilot for *Grey's Anatomy* on my mom's bed," he says proudly. That would have been when he was five or six, a few years after his *Amazing Race* obsession. He and Andy quickly start swapping favorite shows and what's on their current viewing roster. Danny reveals that his favorite show of all time is *Chuck,* the

NBC show that ran for five seasons about a computer geek who gets a literal brain dump of classified information. He recently watched *Mr. Robot*. He and his mom have agreed to binge-watch *The Wire* once he knows where he's going to college.

Danny sits behind Andy and starts pulling up images of mushroom nanoribbons on his phone.

"This is what we're looking for," Danny says, showing a variety of abstract figures. Some look like curlicues of cigarette smoke, others like dry pasta standing on end.

Andy's turning the dials to maneuver around all the parts of the silicon wafer, trying to locate anything that looks like it could be a nanoribbon. He stops when he gets to a busy-looking image of swirls.

"It's almost creepy looking," he says. "It looks like . . ."

"It looks like something out of a horror movie," replies Danny.

Andy keeps turning the dial, moving across the surface of the silicon wafer.

He stops and mutters, "Spongy rods."

Danny leans in and they stare at the monitor.

"To me, this looks like the heart of the matter," says Andy. "I'm getting the warm and fuzzies."

It seems promising, but this is one of those less binary areas of science, subject to some interpretation. Andy wants Danny to have a turn. They switch places and Danny carefully, gingerly starts working the knobs. When they see something they like, they can take a photo they'll burn to a CD.

"The deal with this is, essentially you're looking, well, in this case, you're looking to see what the hell's there," says Andy. "But don't be—how do I couch this—if you've got a range of things there, take it all, because ultimately you're going to take all this data and you're going to sit down and really formulate what's there. So spend some time; it's not unusual for a student to sit here—now William Yin would sit here with one slide for hours, I'm not encouraging that." He laughs.

Danny's a quick study, so he starts capturing images and then loads his other wafers. What he and Andy glean is that the surfaces of the wafers aren't completely covered in mushroom char, which could affect the ability of the battery to function properly.

But Danny stops when he lands on something that looks like a near replica of a nanoribbon shown in the paper. He lights up.

"It appears to be there, so I'm, like, really, really, really happy," says Danny. "I'm trying my best to control myself and not get my hopes too high up. The fact that they exist and even appear to exist, that's good."

Andy's got a fat-cat grin on his face.

After some back and forth and back again, Andy and Danny believe the nanoribbons have materialized. Now Danny can go about constructing his batteries, a relatively straightforward process that involves building a small sandwich of materials to act as anode, cathode, and electrolyte and packaging it in a stainless steel battery case.

All this has taken about twenty minutes, and as Andy and Danny are finishing, a scary teenaged hunger sets in—for both of them. They debate a few dining options and Danny makes an enthusiastic bid for Rare, his favorite burger place, which is a few blocks south of Grand Central. Feeling somewhat triumphant and game, Andy agrees and they hop on the downtown 6 train.

By the time they exit Grand Central Terminal and walk the five blocks south on Lexington Avenue, they're approaching chew-your-arm-off hunger. Danny strides confidently into Rare—it has a saloon vibe, lots of dark wood and dim lighting—and speaks to the host.

You can deduce a lot about a teenager based on how he carries himself in a restaurant when not in the company of his parents. In Danny's case, it seems that being unto the castle born has produced impeccable manners. (It could easily go the other way. It's not hard to imagine kids catapulting spoonfuls of peas and potatoes at each other across the castle's imposing dining room table, which feels like half the length of a bowling lane.) His napkin immediately finds its way to his lap. He never fails to say "please" and "thank you" to the waitstaff. When the food arrives, he doesn't rip into it like a starved wolf, the way the kids do with the pizza. His eyes dart around the table to make sure it's appropriate to start.

He's slightly nervous because he's talked up Rare and Andy's made several joking comments about how this better be a worthwhile burger. No matter what, Danny will likely be redeemed by the truffle fries he's ordered (first asking Andy if it was okay). Most things in life can be forgiven with truffle fries.

When the food comes, Danny seems anxious for Andy's grade of the meal, as if he's waiting for a test to be handed back. It's an easy call because Andy's immediately taken by the array of dipping sauces that arrive with his fries. And the burgers are perfectly cooked and not overdressed. They are ridiculously tasty, memorable burgers. The basic accoutrements of lettuce, tomato, and onion look like they're from central casting for produce.

Not only does Danny relax, he uses the satisfaction at the table to point out that this is actually his domain.

"I know food," he says, with an authoritative nod. "And I know a good hamburger." At this point, as he and Andy are savoring their burgers, well on their way to a good food coma, there's nothing to debate about this statement. And Danny's not merely a consumer—the Slate children engage in fierce cooking competitions with one another. They draw cards for courses (appetizer, entrée, or dessert) and are timed and then rated by the grown-ups, a sort of *Top Chef* meets *Family Feud*. These competitions mostly take place at the Slates' lake compound in New Hampshire. Danny prevailed over the summer with braised short ribs.

Partway through the meal, Danny tells Andy that he and some seniors have a plan, half cooked in reality, half in fantasy, to go on a cross-country road trip the summer after graduation. There's one minor hitch: Danny doesn't have a driver's license.

"My whole family makes fun of me for not having a license, but no one seems to want me to drive," he says slowly, the irony seeming to occur to him at this very moment.

"That's a problem, pal," laughs Andy.

The Slates have employed a nanny for most of Danny's life. As the kids have aged, her duties have morphed and one of her main current functions is driving the kids around town. (Danny seems a bit embarrassed about the whole thing, initially describing her as "the nanny for my brother and sister." Which is true, but he certainly isn't hitchhiking to school. And despite some slight disapproval from his parents, the nanny dutifully delivers the weekly pizza.)

When the check comes, Danny takes out a card and immediately offers to pay, a gesture that registers as earnest and sweet and not at all for display.

"It's my money from my bar mitzvah," he says quickly, appearing to be willing to use his rite of passage money some four years later to treat Andy to hamburgers. He mostly wants it to be clear that he's not dining on mom and dad. Andy naturally declines but acknowledges the kindness of the gesture.

With their bellies full, Andy and Danny step out onto Lexington to make their way to Grand Central to get back to Greenwich.

They walk with no urgency or stress, only ease. It's a walk of complete satisfaction.

8

William

Back on August 11, a few weeks after he and Shobhi texted Andy in the middle of the night, William checked the Google Science Fair site . . . and learned that neither he nor Shobhi had advanced to be named global finalists. He was disappointed because even though GSF had only debuted as a competition a few years ago, it had that Silicon Valley sheen that generated an unusual amount of notoriety, as evidenced by Olivia's megawatt media attention following her win for her Ebola test.

William rarely spends much time dwelling on losses because his life is crammed with too many other things. There were several other science competitions ahead, and of course, he knew that he'd soon be consumed with college applications.

Greenwich students who achieve various things sometimes receive awards at the town's board of education meetings. These moments provide a sanity check to the often embattled, impassioned meetings; the appearance of actual kids serves to remind the adults why they're there, slowing the descent into ugly bureaucratic conflict. Overseeing nearly nine thousand students across eleven elementary schools, three middle schools, and one high school, the Greenwich Board of Education wields sharp claws and a lot of power, sometimes unleashing highly charged decisions, such as the one in 2015 to reinstate a controversial high school music teacher accused of bullying students, despite an arbitration report recommending that his suspension remain in place. The unprecedented decision to ignore the arbitration report undercut then district superintendent Dr. Bill McKersie, as well as GHS headmaster Dr. Chris Winters, both of whom had suspended the teacher after a series of warnings. The superintendent, well liked and viewed as highly competent, left at the end of the 2016 school year.

So when Greenwich students perform, speak, or are acknowledged at the board of education meetings, for a few minutes everyone can set aside their differences (or at least do their seething in silence). At the September meeting, William and Shobhi are being acknowledged for becoming Google regional finalists.

Despite the fact that Mark and Wendy Yin have been attending awards of all descriptions for both William and Verna since their children exited the womb, they still treat every accolade with gravity. So on September 22, the Yins get takeout pizza, William dons a tie and blazer, and they head over to the Central Middle School gym in their rattling minivan for William to get his certificate.

The poorly lit gym is packed with people tonight because the board is voting on one of the most contentious issues in recent years: school start time. Greenwich has been locked in a fiery debate about whether to move the high school start time from 7:30 A.M. to 8:30. Those in favor of the later time cite a body of research regarding the teenaged brain and the need for sleep. Those against say that moving the start time will interfere with after-school activities, sports, busing, etc. Tonight, people on both sides have come out in droves to make a final push, and there's a bristling tension in the atmosphere. People who want the school start time moved to 8:30 have been asked to wear white, and so Wendy is wearing a white polo shirt that she ran to put on before they left the house.

Andy's wearing a suit and has prepared written remarks about William and Shobhi's projects. For him, the later start time would just mean sitting in worse traffic, so he's opposed to the change. Due to his track record, he's a regular at these meetings, but he manages to speak about the kids and their accomplishments with genuine enthusiasm. He takes the podium, introduces his students, and distills their complex projects into a whiz-bang description for the lay audience. Shobhi and William walk up to Andy to applause, collect their certificates, shake hands, and get their photo taken.

Afterward, William lingers in the auditorium. He'd like to get a word, if possible, with Dr. Winters. It's about the college recommendation that William is asking him to write.

"I think it's better if you ask him at school," Mark tells William.

"But I'd really like to talk to him now," says William. He realizes it's a bit of a frenzied moment to ask for a sidebar with the principal. William had been in a freshman discussion group led by Dr. Winters and the two had developed a tight relationship. Winters had a real affection for William, whom he found intriguing on so many levels—so much more nuanced and pithy than he'd expected. He admits that initially he failed to see William's multidimensional nature.

"I put him in a box," says the headmaster. "A stereotypical, hard-working kid who's good in math and science. But then, there's William leaving cookies on my desk at the holidays, there's William playing in the pep band, and there's William with his Bernie Sanders sticker on his laptop." In short order, William showed that there's really no box into which he fits.

William visited the University of Pennsylvania, Harvard, and MIT the summer after his sophomore year. He decided he wanted to experience this rite of passage on his own. Mark and Wendy agreed to let him travel solo, seeing no real reason not to let their son, who asked them for so little, venture off alone. William enjoyed every aspect of it—booking his travel, taking the train, and most of all, going on campus tours. It was liberating to be someplace with zero pressure or baggage. He could disappear, slide into anonymity. And it gave him a brief but sweet sample of the freedom he'd have in college.

There wasn't much in the way of a discussion in the Yin household about where William would apply early. For Mark and Wendy, Harvard stood as a beacon of aspiration and hope, the ultimate symbol of William's intellect and achievements. Wendy often said that Harvard was the only college people back in China had heard of. And it wasn't as if William *didn't* want to go to Harvard. At the same time, he didn't share Wendy's flame-lit yearning.

Given his twenty AP exams, his perfect ACT score of 36, and his performance in Andy's class, Harvard didn't seem to be a reach for William. When he told Ms. Patti of his choice to apply early there, he was met with nothing but encouragement.

The summer after junior year, he'd spent six weeks at MIT's prestigious Research Science Institute (RSI), a highly selective program for the Williams of the world. By its own description, it's for "80 of the world's most accomplished high school students." Andy always said that for kids accepted to RSI, they practically had a ticket to Harvard or MIT. His former student Ryota Ishizuka had gone to RSI and like William had made it to Intel and won a slew of prizes. He'd gone to Harvard.

This year, there are ten Greenwich High kids applying early to Harvard, including four in science research: William, Olivia, Danny, and Henry Dowling, the last two being legacies.

William has been considering very carefully how to present himself on his Harvard application and what to include. He opts not to send a CD of his piano playing.

"I just don't think it's good enough," he says. "I don't think it would really help me." This, coming from a kid who's performed at Carnegie Hall and earned a trip to Portugal for his entire family while he gave concerts there.

Madeleine Zhou, an accomplished violinist who plays in selective orchestras, is ambivalent about submitting a CD with her college applications.

"I'm not that good," she says one day while discussing the heap of colleges she's dealing with. In this moment, as the kids critique themselves and come out with "inadequate," you realize that a perverse thing happens in the Greenwich incubator of überachievement. Being surrounded by peers who are just as brilliant and accomplished as they are leaves them feeling extremely average. All the greatness seems to cancel itself out. Rather than seeing themselves as part of an elite club, they focus only on what they perceive to be the echelon that's even better. It's depressing. And it makes you wonder how kids in this town feel who actually *aren't* the academic royalty.

William decides that the cornerstone of his Harvard application will be the two science projects he did in Andy's class, his targeted cancer therapy and sticker to detect atherosclerosis. Both had cleaned up at various competitions, but more important, William feels they are a precise window into who he is. They shine a light on that Rube Goldberg brain of his, the quality of his thinking, and his willingness to persevere through seemingly indomitable quandaries.

He also has a vague idea about pursuing a new science project with Andy, a rare undertaking for seniors. Research is a kind of mental oasis for William and it's appealing to think he could do one of his favorite things during the college application churn. But he hasn't mapped out any firm ideas. He's got a sense he'd like to do something brain related, a whole new realm for him. For now, though, he has just this tiny marble of an idea rolling around in his head, pinging an array of possibilities: depression, brain mapping, Alzheimer's disease. The truth is, William can't devote the mental real estate to conceiving of a detailed project right now. He has to focus on his essay for Harvard, which he revises again and again, scrapping whole versions and starting fresh.

And so the marble of his idea continues to roll.

9

Sophia

A s soon as Sophia joined the class, she dove into the literature about Lyme disease, poring over papers about the strange ailment, looking at its molecular underpinnings.

One day she came across a paper that explored the notion of biofilm in Lyme disease. Biofilm is a collection of cells that stick to one another to form a visible gooey surface layer around other cells. She read that biofilm may act as a biological wall that surrounds Lyme disease cells, essentially blocking antibiotics from reaching them.

This was an intriguing clue. She kept digging and found that biofilm can also form on medical devices implanted in the body, such as mechanical heart valves or central venous catheters. There was some research suggesting that a class of drugs called PDE4 inhibitors could be effective in degrading the biofilm, thus allowing the antibiotics to reach the cells and do their work.

So she decided to design an experiment where she'd grow her own Lyme biofilm in Andy's lab and administer the PDE4 inhibitors to see if the drugs would destroy it. If successful, she'd add some of the standard antibiotics given to Lyme patients—the ones she herself was taking—and see what effect, if any, that had on the bacteria.

It was an ambitious idea. She'd be dealing with the live *Borrelia* bacteria (albeit a nonpathogenic version) that cause Lyme. She'd need to figure out which PDE4 inhibitors she'd apply to her biofilm and all of her work would need to be validated by using the SEM at Sinai.

Andy loved the project. He loved that Sophia had connected a bunch of scattershot dots to come up with a reasonable hypothesis. And the personal nature of her project made it irresistible. How compelling for a kid to describe her inspiration being her own suffering from a disease? He'd had a few other students do Lyme projects in the past, including Danny, but this one was particularly enterprising. Her project was grounded in a certain logic and, while painstaking, seemed promising. The big unknown, of course: Would it work? Setting aside the precarious business of growing biofilm, there was the wild card issue of whether PDE4 inhibitors would chew it up. A project like this was rife with built-in hurdles, big and small, but Sophia Chow had endured more than her share of obstacles in recent years.

Stephanie and Gregg Chow are California natives who moved their family to the East Coast for Gregg's finance job on September 9, 2001. Following the events of September 11, they considered turning around and heading back to the Bay Area but ultimately decided to stay.

One of Gregg's colleagues turned him on to Greenwich, a place they'd never visited. "It looked like something straight out of a Norman Rockwell painting," says Stephanie. It was a certain kind of New England, old shingles and shutters pretty, a stark contrast from California's stucco and palm tree pretty.

With so many distinct neighborhoods and properties, the Greenwich experience varies widely depending on where you live. For those with the means who desire isolation, you can live in Greenwich's northernmost expanse, known as "the backcountry," behind the walls of a three- to six-acre gated estate and never see your neighbors. Without a full lay of the land, the Chows spent their first year in the backcountry. Stephanie was at home with her twin toddlers, knew no one, and felt removed from humanity.

So they moved to the affluent neighborhood of Riverside, the kind of place where gaggles of kids can trick-or-treat in packs, zigzagging across the streets from one fabulous, spookily decorated house to the next. But because Riverside is home to many moneyed people, it can feel a bit snobby, which is saying something in a town whose baseline greeting is an air kiss.

It wasn't long before Stephanie figured out what she really wanted, which was to live on Greenwich's coveted waterfront. She and Gregg had once owned a lakefront bungalow in California and she'd never stopped regretting that they'd sold it. So as soon as they moved to Riverside, Stephanie started keeping close tabs on Greenwich property along the Long Island Sound, which on the low end goes for about $4 million these days.

In April 2012, the Chows were driving around with their realtor when Stephanie saw a sign in Old Greenwich's exclusive Lucas Point neighborhood for an odd modern home with a flat roof. Stephanie barely cared about the house, its expansive views of the Sound dwarfing any concerns about the structure. Plus, considering the Greenwich bubble of excessive prices for everything, the property was a relative deal.

Still, Gregg was skeptical. "What if there's a hurricane?" Stephanie quickly dismissed this, as she was giddy that the house on West Way was within their reach financially. They made an offer and were thrilled when they got it.

And it was at this house that Sophia found her perfect happiness. When summer came, the Chows launched a dock into the Sound. Sophia would sit on it by herself at night, watching the moon and listening to the water. This and this alone brought Sophia a sense of decompression. It can be hard to find respite from the noise of adolescence, but here it was. It was private and peaceful and hers.

Six months later, following Hurricane Sandy, the Chows wore thick rubber boots and stood in their living room with contractors and insurance adjusters. The Long Island Sound had washed with terrific force right into their home, and after the three and a half feet of water receded, there remained a dense muddy sludge whose stench was so foul, they couldn't stay indoors for more than a few minutes. In many ways, the water damage and mud was worse than if the house had just floated out into the Sound. It was like having your house robbed by a burglar who insists on lingering and defacing all your possessions right before your very eyes.

Though it was still standing, the dream house was gone.

It was traumatic for Sophia and Brandon. They made a brief trip to the house to try to retrieve some belongings. Sophia focused on her dog, Brady, trying to minimize any trauma the shih tzu might have been feeling from all the chaos. The Chows had waited out the storm with friends, like one big family sleepover, thinking they'd return immediately after it passed. When Gregg informed Stephanie of the level of damage, she had one good cry and then told herself that was the extent of the pity party.

The destruction of their home was the start of a two-year odyssey of moving from rental house to rental house (three in total, plus a loaner from a friend) while trying to decide whether to offload the property or hold on to it and build a new house from the ground up, a process that would entail sparring with the Greenwich zoning board, architects, contractors, and neighbors with picayune concerns about god knows what. But they needed to figure it out because they were being jacked on the rental prices, which shot up again and again.

Brandon and Sophia would pack their belongings so many times, it became part of their muscle memory. They were the nomads of Greenwich, albeit in Range Rovers. Every temporary home was well appointed and their needs were well met, but it was deeply unsettling. Kids need a home base and consistency. They need to be able to toss their backpacks, without a thought, into the same spot every day.

Sophia coped with the stress by redoubling her efforts on her schoolwork. Since her home life had spiraled, she focused on the things she could control. But she couldn't stop thinking about Sandy and the damage it wrought—not just to her family but to less fortunate ones too. Despite growing up in the 1 percent bubble, Sophia is a naturally empathetic and self-aware kid who knows she lives a rarefied existence. So one day she went to Gregg and said she wanted to start a charity. He didn't hesitate to help her. Since he calls her Sophie, she chose to name her nonprofit "Sophie Cares." The mission statement she wrote when she was thirteen says:

> In November 2012, I lost my home to Super Storm Sandy. There was over 3 feet of water in my house and not only did I lose my house, but I also lost pictures, keepsakes and school work. Although my parents had insurance to cover the items, I still longed for some things that could make my life feel "normal" again. My backpack, some books, my art pencils, something that could connect me back to my friends again so I could begin my own recovery. We were fortunate because we were able to replace many of these things over time, but I got to thinking about other families and kids who may not be as fortunate.
>
> These kids just want to get back to normal as soon as they can and getting them books, pencils, backpacks, papers, things they can write letters with . . . all help. Trust me, I know firsthand. I started my charity to meet the needs of these kids . . . this is a really important mission. Please help if you can.

Gregg saw the nonprofit as a way to give Sophia an emotional outlet for all the disorder she and Brandon had lived through.

And after many late-night conversations, Gregg and Stephanie decided to stay put. They'd rebuild the house. Gregg saw the whole thing as an important lesson for his kids: When things don't go your way, you don't just walk away. You endure some of the short-term pain because, guess what? Sometimes life is pain. Between her Lyme disease and the family's dream house being wrecked by the elements, this was something Sophia had already learned.

10

Romano

During his sophomore year, Romano and his mom, Kim, traveled to Tucson, where he met a plant expert who gave him a primer on desert plants. Romano was captivated because he'd always been drawn to horticulture and natural remedies—a great source of mockery from his siblings.

"The kid has an herb garden," says Sophia Orlando, eighteen, pausing to let this sink in. "You didn't know about this?"

Sitting in the Orlando living room while briefly home from college, Sophia delights in dishing details about her little brother, with whom she shares a motherly-sisterly dynamic. When she was in high school, they frequently had long talks into the night, discussing every aspect of their lives. Now that Sophia's in college in Arizona, they speak on the phone for extended stretches at least once a week. The Orlandos have never embraced texting for intrafamily communication. They pick up their phones and actually speak to one another.

Romano keeps a wooden planter at the back of the house where he grows zucchini, eggplant, jalapeños, sweet peppers, and three different kinds of herbs, which he gives to Nanni, his father's mother, for her homemade sauce. The garden is in keeping with what Sophia and their oldest brother, Dario, say are Romano's "old man" behaviors. He likes to have a glass of red wine at night while he does his home-work at the dining room table, a habit his parents sanction as part of their son's young fogey behaviors. Plus, in an Italian household, a glass of red wine is practically compulsory. And during the summer, his morning ritual involves a cappuccino and biscotti. His friends call him "Grandpa." But to his siblings, the garden took their broth-er's quirks to new levels.

The desert plant guy told Romano about *Jatropha cuneata* and said that among some Native Americans, the plant is known as the blood of Christ because of its healing properties for the skin. Romano was rapt as he watched the guy snap open a piece of the plant, prompting red droplets to trickle out.

Several months later, back in Greenwich, Romano was officially accepted into Andy's class and needed to come up with a research project. He wondered if he might be able to work with *Jatropha cuneata* to make some kind of healing elixir for the skin, something that could adhere to a wound.

During summer break, Romano texted Andy, who was glad to see the significant upgrade in thinking but had to break the news that it would be impossible to simulate in Connecticut the desertlike condi-tions necessary to grow or even maintain the plant.

Andy gave Romano an assignment of sorts. He told him to read *Science Daily* each morning and see if there were any papers or topics that lit his fire. For students who needed some guidance on how to think about a project to which they could fully devote themselves for a year, Andy told them to start by peering under the scientific re-search paper kimono. These papers weren't necessarily published in major journals, and many reported on incremental findings or deeply weedy stuff, but there was such variety, just combing through was useful. This also showed the kids the template for a research paper, which their posters for the competitions would emulate.

One day over the summer, as Romano sipped his cappuccino and nibbled his biscotti, he found a paper about an amino acid called L-DOPA, which is found in the hairlike threads of mussels. Naturally sticky, L-DOPA is what allows mussels to adhere to rocks.

Romano paused when he saw the article. It was instantly intriguing to him because in keeping with his garden, he liked the idea of working with a natural substance. Even though the desert plant option was dead, he wondered if L-DOPA held any promise. Because it's a natural adhesive, maybe he could set about fashioning a squirtable liquid bandage that would harden once it was on the surface of the skin. And maybe he could even add in some bonus medicinal aspects, similar to the compounds in something like Neosporin.

He didn't like it, he loved it.

Andy had had kids work on so-called smart bandages before, but none had ever tried to create a liquid bandage that would solidify on a wound. Andy thought the idea seemed cool. But what Andy liked more than anything was that Romano finally had purpose for the class.

Romano got in touch with Bruce Lee, Ph.D., an authority on bioadhesives. He couldn't help but smile and titter every time he mentioned the guy's name and he could never not say his name in full. "I'm talking to Bruce Lee from Michigan Tech. That's hilarious. His name is Bruce Lee. Like, that's so weird."

Romano had a series of lengthy phone chats about L-DOPA with Bruce Lee, who was apparently very happy to see a high school student delving into his area of science. No one had ever made a liquid human bandage using the amino acid before, which could be viewed as either incredibly original or impossible.

By the beginning of October, Romano is working on a research plan. He's feeling good; he likes the novel simplicity of his project. Liquid bandage. And he isn't missing football in the slightest. He sometimes goes to his best friend's house on Friday afternoons, and as he walks past the football fields, he's pleasantly surprised at the absence of longing. On the contrary, he says, "There's just this joy I'm not there."

But he's weighed down by other, more personal matters.

Romano and Caitlin have been dating for seven and a half months and he's starting to get the gotta-go itch. Caitlin has been his first real girlfriend, and with no awareness that he's talking straight from the breakup playbook, he says he'll "always be glad" they were an item, that he has "no regrets," and that he "wouldn't trade the relationship for anything." All of this rolls out of his mouth with such ease, you wonder if male DNA is encoded with such language, whether evolution has preserved it as a trait somehow key to the survival of the species.

There's no major event driving his desire to part ways, but he feels like the relationship has run its course. Unfortunately, Caitlin's initial confidence in landing Romano as her boyfriend didn't hold up. She's dogged by her social group inferiority, which plays out in weird ways. He gets invited to the popular-crowd parties and she doesn't, so they have to engage in long discussions about how such matters will be handled. It's a lose-lose situation that only sets them down the one-way road to splitsville.

When he talks about Caitlin now, as opposed to just a month ago, his tone is already laced with nostalgia. He's moved on, but he's thinking carefully about how best to handle it. He knows that it 100 percent has to be done in person. None of this breakup via text bullshit. And no phone dumping, either.

While he's mulling his escape hatch options, one day he hears his name over the PA, misses the first part of the announcement, and thinks he has to report to the main office. He walks into his comedy and improv class and says he has to go, but people are eager to correct him.

"No, no! You got homecoming prince!"

Among one of the stranger GHS traditions is the nomination of the homecoming court by the teachers. The student body then votes. This is Andy's twelfth year at GHS and until now the number of homecoming nominations he's cast has been zero. But he liked the idea of a science research kid possibly landing the title and Romano was an obvious choice. This was Andy's doing. Romano later learns that another teacher also nominated him.

When Romano walks into the lab that day, he's greeted with hearty congratulations, a bit of awe, and in a turn both sweet and surprising, zero sneering. It's easy to imagine kids who don't play in the popular sandbox wanting to pee in it, but they seem almost proud that Romano has become one of them without sacrificing his lofty perch in the social hierarchy.

Romano is a winning combination of grateful and slightly embarrassed. "Thanks, guys, thank you," he says, grinning while looking down.

"Congratulations," says Andy, with a big knowing look.

"Oh, Mr. B. Thank you so much."

There's just one wrinkle: Caitlin is homecoming princess. She's euphoric and Romano says she's been dreaming of this accolade "since childhood." This means there can be no breakup until after the weekend, which includes the full fall smorgasbord of events: pep rally, dance, parade, and then, of course, the actual half-time coronation. You can imagine the school newspaper headlines if he does the deed too soon: "Prince Shatters Princess's Dreams; Hasn't Been Seen Since Breakup . . ."

It'll have to wait.

11

Andy

The year isn't off to a terribly good start.

Even though Andy navigates the whole competition aspect of his class with extreme delicacy, not wanting to send the message that the only value to the whole thing is competing, there's an occasional loss he finds genuinely deflating, a series of pinpricks to the soul. This fall, he's nursing two losses, one being his own.

Since becoming a teacher, he's learned not to spiral into a whirlpool of despair over the most basic of lessons, which is that life is many things, but fair isn't one of them. And in his ecosystem, kids get heaping spoonfuls of unfair because inherent in this competitive sphere (or any other rigorous competition) is getting overlooked, beaten, and deprived of well-deserved glory.

Such was the case with William and the Google Science Fair. Back in early August, Google had contacted Andy asking him to elaborate about William. When the email landed, Andy felt a certain carbonated giddiness. "Last year, I got the same thing for Olivia, the email asking for more information," he said. He decided to set aside a chunk of time, bring his laptop to a café, and craft a heartfelt letter about his prized student.

Because it was the summer, he had the mental reserves to approach such a thing. During the year, Andy's job came with a suffocating amount of paperwork. Every application to any fair or competition required a highly detailed and technical letter of recommendation. And since his kids accomplished so much in his class, Mr. B. was often the first stop on the college recommendation train. Andy dreaded the deluge of writing each fall. But since it was August and the slag heap seemed far off, he actually took pleasure in penning an ode to William. He wanted Google to know so many things about this kid: his tenacity, his multidimensional personality, how, really, when it came down to levels of vision and ability to execute, William was at the very top.

In response to the questions, he wrote:

I'd also like to point out that William's project was filled with many seemingly insurmountable obstacles, which I believe highlights William's perseverance and resolve, to ultimately create a simple, portable, and inexpensive iontophoretic patch for the detection of atherosclerosis. Somewhat early in the research process, William needed to construct a custom-made vertical diffusion cell . . . I recall seeing William in the school hallway, on his cell phone, at least once a day for a period of 1.5–2 months, as he pleaded with local model-shops to fabricate his diffusion cell at minimal cost, in a timeframe that would serve his purpose. I imagine that few, if any, of these machinists actually said "no" to William's request, however none would give the project the importance and timely response that was needed. Undaunted, William procured the raw plastic material himself from a local plastics warehouse, and worked with the technology teacher at the high school to fabricate the diffusion cell himself, with the help of a techEd student, and the rest is history, as they say.

Also, I'd like to point out that William is a highly skilled and accomplished pianist. I mention this, as his time commitment towards music is often lengthy, and I remain amazed at how he is able to juggle his commitments to put his best foot forward in all that he does . . . academics, music, extracurricular clubs, and his atherosclerosis research.

I apologize for being somewhat long-winded . . . I am so passionate about all of my amazing research students, and take every opportunity to describe just how incredible they are!! Honestly, however, William is one of the most incredible students I have ever had the pleasure to mentor.

Take care,

Andy Bramante

The thing was, no matter how much William and his tortured work habits could detonate Andy's patience, it was almost impossible not to be inched into the kid's corner one long lab night at a time. There was his dimpled man-boy grin, his hair that was something like combed but other times pure bedhead, his always soft voice. And as the son of Chinese immigrants, whose life choices were made in direct proportion to their desire for their kids—more than themselves—to succeed, William's mission was anchored by the whole American dream thing.

Andy saw and felt all of this and he wanted badly for William to continue on his winning trajectory for his plaque sticker.

So when Google posted the global finalists on its fair site and Andy saw William wasn't one of them, it was as if someone tossed battery acid on his heart. Shobhi also didn't advance, and while that gave an extra splash of pain, as a junior, she'd have many more opportunities. Not so for William.

And then there was the issue of Andy's own failure to win a prize he admittedly would have loved to have. Ray Hamilton, Andy's Greenwich High science research predecessor, and Wade Elmer, a Connecticut state scientist, nominated Andy for the National Science Foundation's Presidential Awards for Excellence in Mathematics and

Science Teaching, the most prestigious teaching award for STEM in the United States. Two teachers are named from each state. In 2015, Andy was one of four finalists in Connecticut. The application and selection process was intense, with judges visiting Andy's lab. He'd had a bit of an odd feeling about it, worried whether what he did would resonate with them since the class didn't fall into the traditional models of classroom teaching.

When he found out that he didn't win, he was gutted. It was a rare moment of Andy wanting a distinction he felt he'd genuinely earned—and deserved. In 2012, he was named Greenwich's Teacher of the Year, but the Presidential Award felt so much bigger.

Now in September, he gets the judges' feedback and it's nothing short of maddening. For one, they dinged him for somehow improperly formatting the application form essays. Half of one of his essays was missing and instead of requesting it in full, cutting him some slack for the innocent oversight, the judges held it against him. Other comments seemed to suggest that there was no way to show how effective Andy is as a teacher, since there are no tests and no curriculum in his class. There was some notable praise, too, but like anyone getting feedback after a loss, Andy seems to only have eyes for his alleged shortcomings.

"You know, I was worried that they wouldn't get it, and they didn't get it," says Andy, sitting at one of the low soapstone lab tables. "I'm walking them around, showing them all the instruments and I was asking myself, Are they going to be able to really see what it is I do?"

Another thing that's dragging him down is his perennial stress about money to run the class. Every year, the school gives him $1,200 for science research. With this year's crop of forty-eight students, that works out to $25 a kid. A single project can run in the thousands. He asks the kids to foot the first $150 of their projects, but after that, the class pays the rest. Long ago, he figured out that if kids paid the full cost of their work, it would give those with means an unfair advantage.

Some of the competitions give prize money to winning schools, and due to his track record, Andy almost always gets a few thousand bucks there. But with the help of some parents, every year he holds a very casual afternoon fund-raiser at the Bruce Museum Seaside Center at Tod's Point, where parents eat sandwiches and he does his dog and pony show about the class. He explains how little money he gets from the school and gently solicits donations. He makes it clear that whatever money people contribute is not earmarked for their kid, that they're contributing to the fund at large. Some years, he can collect several thousand dollars, but other times, very little.

He's made many pleas to the school to underwrite his program, but he's been consistently rejected. The answer from the administration is always the same: get the parents to pay. One year, he requested funding for twelve new laptops for the lab (many of the instruments run with laptops) and he was turned down. He paid for them out of his own pocket and later reimbursed himself from the parent fund. He's also brought in what he estimates is about $200,000 worth of scientific instruments—freebies he's gotten from his industry cronies. This kind of equipment is rarely seen in high school labs, apart from STEM schools that are devoted wholly to science.

What he dreads every year is the ask. He can't stand shaking his tin cup and groveling for money to keep the program running, especially because it's an official honors course offering, not an after-school club. Having to plead for money from parents feels tacky, especially given all the class has accomplished. Every year, he loathes doing it, and yet, every year, he does it.

When Andy walked away from two decades as an analytical chemist in corporate America, there was no way for him to know that working with high school kids would both wring him out and raise him up the way it does, that he'd enter an emotional and mental terrain he'd never touched while climbing the company ladder.

In 2005, when he finally broke from industry life, he'd badly needed to make a change—for many reasons. But scrapping his successful career was a radical move for a working-class guy from the Bronx who'd made it out.

When Andy was pumping gas during high school and college at Dallacco Service Center, a family-run station in the Woodlawn neighborhood of the Bronx, he was paid a flat fee of twenty-five dollars a day, no matter how many tanks he filled or how many tires he changed. But there was one opportunity to make more than double the day rate. Dallacco had a small tow lot piled with cars in line for resuscitation. Though it was fenced in, the cars were easily accessible and vulnerable to theft or vandalism. The owners of the shop paid Andy an extra thirty dollars to sleep overnight in an un-air-conditioned metal trailer in the corner of the lot. This passed as security in the Bronx in the 1970s.

Andy jumped at the extra thirty bucks. "Sure! I'll sleep in the trailer," he recalls with a laugh. "I mean, thirty dollars back then was a lot of money." Sweetening the deal, the trailer was equipped with a thirteen-inch black-and-white television with rabbit ears. So he could do his homework *and* have a TV all to his lonesome. Talk about living the dream. Just three nights and he could pay for the paint job on the 1972 Chevelle his grandfather and namesake had bought for him. For sixty-nine dollars, he transformed his car from a garish gold to a cool dark blue.

He worked at Dallacco all through college since he was still living at home. Being raised by a single mom who slept on a pullout sofa in the living room of their apartment so he and his older brother, Jerry, could have the only bedroom, Andy chafed at the idea of starting college actively digging his own crater of debt. That went against every aspect of his family's ethos. His grandparents, on whom his mom leaned hard, were of the cash-and-carry generation and never used credit cards or took on debt in any form.

By both upbringing and DNA, Andy was built to hustle. He got his first job at age fifteen, the minute he had working papers, selling ladies' shoes at Thom McAn. His mother, Catherine, eventually worked as a pharmacy technician at the local hospital, just steps from their building on East 232nd Street, but not before a series of jobs that included driving a limo, taxi dispatching, selling insurance, cleaning other people's apartments, and eventually cleaning the Catholic school attended by Andy and Jerry.

His mother and grandparents formed a parental triangle around the boys. His grandfather, Andy Cantiello, was a quiet, retreating type who worked as a city custodian. He made the most of his pension and eventually bought his dream house in the Poconos, only to die of lung cancer at age sixty-eight. He was soft but not demonstrative, and it was from others that Andy and Jerry would hear that their grandfather spoke nonstop with geyser-like pride about his grandsons.

The elder Andy wasn't capable of working solely one job; he had the full-on son of immigrants grind, where you take nothing for granted, assume the worst, and work your ass off as a way of trying to head off every possible life calamity. On weekends, he oversaw cleaning crews for stadiums and concert venues, often enlisting his grandsons.

As for Andy's father, he was out of the picture when Catherine was pregnant with her third child, who would be stillborn when Catherine was seven months along. Andy's father had been running around and Catherine, not going to take another minute of deceit, turned up at the home of the mistress du jour and told her husband to make a choice. He chose the other woman, leaving Catherine with two boys under the age of four. She briefly went on public assistance.

Andy's father would drift in and out of their lives without any consistency, seemingly unable to crawl out from under the deadbeat dad boulder. Unfulfilled promises to show up, forgotten birthdays, and nothing in the way of grand gestures to make up for his many lapses—he was reliable only as a source of steady disappointment. So Andy, Jerry, Catherine, and her parents forged ahead without him. The Cantiellos paid for the boys to attend Catholic school. Andy was flagged early on as a bright kid and skipped third grade. There was never a question about whether Andy would go to college, so when he got a free ride to Fordham University, right in the Bronx, it was the logical choice. He earned a degree in chemistry and then stayed on for a master's before propelling himself out of the old neighborhood, but not before he worked temporarily as a UPS driver to make money before landing a decent job.

That Andy now lives in a quaint neighborhood in Fairfield, Connecticut, and spends his days in Greenwich marks a life trajectory that was hoped for but in no way guaranteed—the product of aspiration, sweat, and the occasional stroke of luck. He still hustles, running his side business repairing scientific instruments. The working-class Bronx in him will almost never let go an opportunity to make some cash.

But it's that same go-go instinct that gives him the muscle to keep hanging in and on with these kids, even when the cost-benefit doesn't seem to be working in his favor, like this fall.

On October 19, the Siemens Competition results are posted online. Of Andy's dozen or so kids who applied, just one is named a semifinalist: Olivia Hallisey. Her Ebola test just keeps winning. Great news for Olivia, but another sock in the gut for her teacher.

12

Olivia

When Olivia Hallisey was a freshman, she watched with horror as the Ebola epidemic ravaged West Africa. Between the loss of life, the attempts to combat a disease against a grim and despairing backdrop, and the world's stomach-turning apathy, the Ebola epidemic was a particularly ugly and shameful tragedy being played out before a global audience that had largely nodded off.

Not so for Olivia. She had been attuned to the power of medicine from birth. Her maternal grandfather, Dr. Harold Kosasky, was a notable Boston ob-gyn and infertility researcher. As a child, Olivia loved nothing more than going to work with him. While his towering career left a deep impression on the entire family, it was perhaps felt most deeply by his first granddaughter.

As she kept apace of the Ebola epidemic, one day she saw the news that an American physician who was working in West Africa as part of a Christian mission had contracted the disease. Dr. Kent Brantly was going to be sent back to the United States for treatment, a move that escalated the American public's panic to irrational levels. Despite the attempts on the part of U.S. public health officials to explain that Ebola was not airborne and could only be contracted through close contact with bodily fluids, people were besieged with needless hysteria—while thousands lay dying in West Africa.

In that moment, Olivia knew she wanted to do something Ebola related for her project in Mr. B.'s class. Olivia was new to the Greenwich public school system. Up until high school, she had attended Greenwich Academy, the tony all-girls private school in town. But she wasn't coming to Andy's class totally cold. Her two older brothers, Will and Blake, had done research during Andy's earliest days teaching at GHS. The Halliseys have long been big supporters of Andy.

Olivia started reading about everything from diagnosis to treatment—and learned that diagnosis of Ebola was costly, complex, and not well suited for the regions where the disease was most likely to take hold. The main challenge was that the reagents (chemicals) required refrigeration, which is hard to come by in rural West Africa. This barrier gave the disease a great advantage since it was impossible to perform rapid, widespread testing.

So Olivia decided she wanted to design a cheap, fast diagnostic test that didn't require refrigeration.

She knew her test would be ELISA based, meaning it would use a bodily sample to test for antigens that, if present, would produce a visible color change, such as in a pregnancy test. The question was how to make a test that worked at any temperature.

Olivia embarked on a grueling, often frustrating odyssey to crack this problem. She encountered numerous hurdles and professional skeptics who condescendingly and sometimes competitively told her and Andy they were destined to fail.

While trying to obtain an ELISA kit, Andy called one company that flatly told him, "You have no business doing this with a high school student," to which he replied, "Let me be the judge of that." Andy found such interactions patronizing and infuriating. And as a former chemist, it was a face slap of a reminder that he was no longer on that side of the equation—and that some people viewed what he and his kids do as mere child's play.

As for Olivia, it was her first (but far from last) brush with a scientific world that can be harsh, competitive, sexist, and uncharitable. She and Andy eventually found someone who would sell them the ELISA kit. "We got it from Dennis, a guy who distributed it. He thought we were nuts," says Olivia. Though she never met Dennis, after so many calls to companies who refused them, he's emblazoned in her mind as The Guy who yielded to her request and helped launch her project.

Her learning curve was steep. When her research suggested that silk might be the stabilizing force that would alleviate the need for refrigeration, she traveled to the Tufts University lab of Fiorenzo Omenetto, a renowned biomedical engineer, to learn how to make liquid silk by boiling the cocoons of silkworms, a painstaking and time-intensive process.

Olivia was right. Silk turned out to be the key ingredient in her project because of its durability and versatility. And for her purposes, its main asset was its stabilizing force, which previously had been rendered by refrigeration. She processed silk into liquid form and then dipped into it paper that used ELISA technology. If you dropped a blood serum sample containing antigens for Ebola onto her paper test, it changed from white to blue to yellow.

Ultimately, over the many hours and late nights in the lab, her frustrations and failures turned a corner and she started to see success. At CSEF in 2015, she handily won her category and a trip to Intel ISEF, where her winning streak continued. She received a first-place award in the biomedical and health sciences category, arguably one of the most competitive categories at the fair. With that prize, plus a couple of specialty awards, she went home with $6,500.

Six months later she was sitting at the Google awards banquet in Mountain View, California, surrounded by Silicon Valley luminaries of the present and future. At all of these awards ceremonies, they announce the lower place winners at the start and then work their way up to the grand winner. As the ceremony wore on and her name wasn't called, Olivia figured she didn't win anything—and she was okay with that. The whole Google Science Fair experience had been so dazzling. The competitors were treated like rock stars. She felt lucky just to be there.

So when she was crowned the top winner—a prize that included that fifty grand to be put toward college—she was genuinely stunned. As she made her way to the stage, confetti and balloons rained down and her face bore a look of awe. When she was handed the LEGO trophy, that look changed to pure joy.

What happened to Olivia in the wake of the Google win was in some ways even more astonishing. Following the major science competitions, such as Regeneron STS and Intel ISEF, the top winners often get a brief spin on the media merry-go-round. Their local newspapers and TV stations might interview them. A winner makes for a fun, inspiring piece about a brainy kid who's unraveled some scientific mystery and won a lot of money.

But when Olivia conquered the Google Science Fair, the media and the public seized on her and her accomplishment with an unprecedented fervor. She went on *The Late Show with Stephen Colbert*. She was featured by CNBC. She appeared in *Teen Vogue*. She was named to *Time* magazine's 2015 list of Most Influential Teens and *Crain's* 2016 20 Under 20 list, and was written up in dozens of other publications. This September, she's on the cover of *Greenwich Magazine*'s "Ten Teens to Watch" issue. She's got a Wikipedia page, and if you type her name into Google, 50,700 results turn up.

She's also become an in-demand public speaker, delivering speeches in France, Toronto, San Diego, and Arizona.

And in one of her glitzier turns, she was chosen as a debutante for the 2015 Bal des Débutantes in Paris, the ultimate coming-out for girls in high society.

Her parents, Bill and Julia, sent Olivia to the affair with her older brother, Will. Olivia told the magazine *Hong Kong Tatler*, "My parents really wanted me to go. They were like, 'Do it. Be a princess for a weekend.' But there are lots of aspects I like about it. I have loved meeting an international community of people and I've made some really great friends."

While she was in Paris, she was glamorously styled in a Giambattista Valli gown, a sky-blue dress with a fitted short-sleeved top that descended into a cascade of blue feathers. The French glossy magazine *Paris Match* ran a lengthy feature about Olivia, pegged to the ball.

But even a turn being princess couldn't or wouldn't interfere with Olivia's other big competitive passion: swimming. The article noted that in keeping with her intense swimming regime, she trained in a pool in Paris for two hours each day.

Olivia has become a supernova in a way that no one else on the competitive science fair circuit ever has. There's no question that her project was innovative and timely—and that she developed a solution to something that had so furiously hammered the world's panic button. The idea that a clever, determined kid from suburban Connecticut could remedy part of the problem proved too tantalizing for the public to allow Olivia to be just the usual media blip.

Plus, Olivia is a girl in what remains a male-dominated field. Certainly, the field is closer to being level than it was in years past, but overall, science is largely still a boys' club. (Male scientists tend to deny this while female scientists wholeheartedly stand by it.)

Finally, Olivia brought the added benefit of being an anti-nerd. Between her competitive swimming, her polished Greenwich upbringing, and her articulate and quick manner, she's not so much a young everywoman—she's the young woman many people would love for their daughters to be. And with her J.Crew good looks, she's a catalog-perfect aspirational figure: the bright, attractive, well-spoken science star who can outswim you in nearly every stroke—and who just happened to develop a quick and easy test for a terrifying disease that caused more than eleven thousand deaths in West Africa and was threatening to spread to America.

13

Danny

On Monday, September 26, Danny is sitting in the lab, laptop open, his gaze steady and focused as if he's undergoing a retinal scan. You imagine he's crafting something of deep import, as he's walled himself off from the lab chatter. You can practically smell the burn of his intensity.

He says aloud to no one, or actually, everyone, "A type of Oldsmobile, four letters, begins with E . . ." He's working on the *New York Times* crossword, and since it's Monday, the easiest day of the puzzle week, he should have had this wrapped up minutes ago. (He times himself.)

In the five days since he and Andy went to Sinai, Danny's plan of action has been to bake more mushrooms onto the silicon wafers to fully cover the surface. The first step is purchasing the mushrooms, which has yet to happen.

But the following day, Danny walks in with a ShopRite bag containing portobello mushrooms and gets to work. Using an X-Acto knife, he delicately peels a thin layer from the top of the mushroom. He's donned the full lab attire: goggles, lab apron, and slack latex gloves that seem rather cumbersome. Beneath the plastic apron, he's wearing his uniform of pale blue Vineyard Vines polo shirt and khaki shorts. (Danny has two basic outfits: khaki trousers and polo or T-shirts in winter, khaki shorts and polo or T-shirts in summer.) His dark brown hair is tidier than that of most of the boys in class. He's slight in stature with a wide turnout, like his feet want to walk off in opposite directions. He says he has a lisp, though it's so slight you might not even notice it. He does have a separate speech impediment, also very mild, whereby some words seem to get stuck at the back of his mouth. He was diagnosed with sleep apnea as a toddler and he wonders if the tonsillotomy might have caused his distinct speech.

He's taken apart a flashlight and is using the cylindrical barrel to trace twenty-millimeter disks out of the mushroom skin. In short order, a battalion of little mushroom buttons has appeared.

He's also trying to remove one of his precious silicon wafers from a stage, the little device used to insert a sample into the electron microscope. (Imagine trying to separate a round tabletop from its chunky, wide circular base.) The aim is to do so without cracking it.

"Mr. B., how am I supposed to get this off?" Danny asks.

Andy walks over and holds the stage up to get a closer look. No matter how delicate the touch, the silicon wafers are as brittle as stale crackers and extremely vulnerable to snapping—which is exactly what happens when Andy tries to separate the wafer from the stainless steel stage.

"That's how," says Andy.

They both knew there was a high likelihood of this happening. Danny takes the wafer shards, as well as the intact ones that weren't attached to stages, and proceeds to fire up the lab's high-temperature oven, which lives inside a fume hood. Once heated, a campfire-like scent fills the lab, prompting one of the kids to ask if someone's roasting marshmallows. Danny will now have several silicon disks with baked-on portobellos. The problem is that the disks are thirty millimeters in diameter, about the size of a quarter, and he can't locate battery casings for this dimension. He doesn't want to use smaller casings because that would mean having to further snap the silicon wafers, which could compromise the integrity of any battery power.

Danny decides to go to the source, as it were: Panasonic. But he takes his time getting there. He's been consumed with his early application to Harvard, math team, and his persistent habit of staying up until two thirty watching television.

"One in the morning? That's child stuff," he says about his sleep refusal prowess.

His goal is to complete his mushroom batteries in time to apply to Regeneron, the seniors-only competition. The first batch of application materials is due in early November. All of Andy's other seniors are applying with projects they completed their junior year. Not Danny, who's off to a decent start but a long way from the finish line.

A few weeks later, on October 17, Danny spends some time on the Panasonic website and locates the casings he needs, BR3032s, in what appears to be a Japanese section of the site. He decides he's going to call Panasonic and ask if they'll sell him some empty casings in the thirty-millimeter size. He phones Japan from the castle, but as soon as he opens his mouth and starts speaking English, they redirect him to the United States, providing him with a number.

The following day, Danny's in the lab, ready to take another run at Panasonic. He retreats to the smaller room across the hall for some quiet.

"Hi," he says when he reaches a live human. "I'm looking to place an order for a battery . . . that's just it. I can't find it on the website. I'm looking for a specific part of that battery, the top and bottom, without them being connected."

There's a pause.

"For research purposes," Danny says, with faux confidence, practically shooting double air pistols to cap this statement.

The request is unconventional enough to trigger a transfer. Danny launches anew into his spiel.

"Hi. I'd like to place a kind of order for a type of battery, the BR3032. But I don't need the entire battery itself. I need just the shell of it, not, kind of, what's inside. Is there anyone I could talk to to try to see if I could get that part of it? I mean, you're the only company that makes 3032s, so I was hoping maybe I could talk to somebody to see if they could sell the individual pieces of the battery itself and not the entire thing?"

To his credit, Danny is handling the vagaries of telephone customer service rather deftly. He doesn't sound hesitant or unconfident.

"Research. I'm doing a research project where I need the shell itself, but not any other parts of it. Because it needs to fit—"

Pause for an interjection.

"I'm making a battery using mushroom nanoribbons."

This sounds just weird enough to seem true. Danny says it with such an even tone of complete normalcy, it bears no traces of being a stupid prank. "I'll need to probably seal it in a slightly different way," he continues. "Someone else did a similar project and they were able to seal it."

He should probably stop talking at this point. His voice has kicked up half an octave as he furthers his explanation, and he's regressing a bit to his actual age. This may have caught the attention of the customer service rep.

"Uh, Greenwich High School."

The rep accepts this revelation with no further questions and instructs Danny to write an email detailing his request. The rep says he'll see if it's possible that the battery factory can set aside a few casings for Danny's mushroom batteries before the batteries are sealed up.

Danny's quite pleased with himself. He walks back into the main lab and tells Andy that Panasonic might actually help him out, that he's going to craft an email request to put things in motion.

"Good stuff," says Andy.

— — — — —

Nine days later, nothing much has happened. The Regeneron deadline is two weeks away. Danny wrote to Panasonic, but they haven't responded.

It's an "E" day, or pizza day. Danny has banished his good friend and fellow research student Henry Dowling from partaking because Henry failed to get a meatball sub for Danny the day before.

Henry watches the others devour the pizza in rather savage and disgusting fashion. Andy, a neatnik whose fastidiousness extends to every inch of the lab, can't bear to watch the pizza grease soil the tops of the lab tables, so in a flash, he's brandishing the paper towels and Windex.

Being in pizza exile seems like an opportune time for Henry to needle Danny about the state of his project.

"Danny, are you going to apply to Regeneron? Are you going to make it?" Henry says. "The deadline is soon, you know."

"Henry, you think I don't know this?" says Danny between bites.

"Someone has to act like the parent around here and keep you on track," replies Henry.

"Where's Wesley when you need him?" says Danny, glancing around the room for Wesley Heim. Wesley, a junior and master engineer, kindly offered to help Danny pry open other battery casings in the event Panasonic won't sell him the ones he needs.

Wesley is the kind of kid who builds crossbows in the Swiss Alps in the summertime for fun, who has his own 3-D printer and designed and printed a prosthetic hand for a toddler in need. So two days earlier, he came to the lab armed with his personal tools, including industrial-strength pliers and crampers, to set about helping Danny separate an assembled battery.

Danny derived great pleasure from the challenge and Andy stepped back and let them have at it. They opened it with brute force, mangling the casing in such a way that it would not be reusable. Today Danny has taken on the task himself, though he's a bit lost without Wesley. Andy walks over to survey the scene. Danny has on thick gloves and is trying to pry open the casing. Nothing about his approach suggests success. He looks rather awkward, and as if he might be a hazard to himself and others.

"Danny, what are you doing?" says Andy. "I'm afraid for you."

Andy suggests that they try to cut the wafers to fit into a smaller battery casing. Yes, there's the issue of the integrity of the energy, but Andy's looking at all options.

"We're going to go for broke here, dude," says Andy.

"Today's a make or break day, I guess," says Danny. "Quite literally."

Andy wonders if they can score the wafer—that is, use a knife to make an incision that will weaken it and allow them to snap it cleanly into two pieces instead of the whole thing crumbling into several shards.

Not wanting Danny to slice his fingers off, Andy takes out an X-Acto.

"No way to score it," says Andy. "It slides right over. Shit."

"Remember that time we opened the battery with the drill?" says Danny.

Andy now has a pair of pliers and is prying the edges of the battery, which has clearly been designed to resist such a thing. Eventually, after some grunting and groaning, he manages to pry the sealed casing apart.

He and Danny stand there, stunned. They now have a separated battery casing. They quickly look over the gutted battery, trying to deduce the various layers of the insides, seeing where exactly a silicon wafer with baked-on mushrooms might go.

14

Andy

In an apocalyptic scenario in which civilization must resort to cannibalism, Greenwich isn't your town. Very little in the way of body fat, lots of jutting bones. One mother said that during the planning of a school fund-raiser there was a call to forgo bread.

However, if during said cannibalistic circumstance you choose your prey based on pure beauty, you'll be lingering over your options. This is a town of very attractive people. The men all seem to have square shoulders and authoritative jawlines; the women, rootless dye jobs and perfect blowouts. A few generations of this kind of breeding yields a gorgeous but unsurprising gene pool. Many people here look like first cousins.

Nowhere is this more apparent than on Greenwich Avenue, known in town simply as "the Ave." It's the main commercial drag, and with its string of luxury shops and a Tesla dealership, it rivals Manhattan's Madison Avenue. It's a scene, one where you feel like you should be appropriately curated to match the posh backdrop.

Andy makes a rare trip to the Ave. in mid-October in search of an anniversary present for Tommasina. It's a particularly special anniversary, their twenty-fifth. Andy has been devoting a lot of time and thought to coming up with a memorable gift. So it is that one sunny afternoon after school he even pays a visit to the Tiffany & Co. on the Ave., magazine ad in hand for a sapphire ring. In person, the ring is small and understated with an overstated price tag. Standing there comparing the ad with the actual ring, a look of deep Bronx skepticism crinkles his face, a silent *Are you fucking kidding me?*

He passes. They eventually decide to reset Tommasina's original engagement ring, giving it some bonus bling befitting twenty-five years of marriage.

Another high: he and Tommasina pay off their home, which they bought in 1993. They've also saved enough to pay for their daughter Sofia's college education so that she'll graduate debt-free.

Meanwhile Andy himself has gone back to school. He's in his second year doing the coursework at Sacred Heart University in Fairfield to get his 092, the administrator's certification that would allow him to apply for jobs like department head, assistant principal, or even principal. (He was also an adjunct professor at the university, teaching night school chemistry during his corporate life.) He drives to class on Tuesday nights and sits through a few hours of lecture. He actually enjoys aspects of it, mainly the parts that give him ideas about future job options. The Sacred Heart class is one of the more steadying aspects to his life right now.

On the downside, the lowest moment of the fall is the loss of the Bramante family dog, a bichon they adopted as a puppy when Sofia was six years old. She named it The Cuteness and he became a sort-of surrogate son, the fourth member of the family. When Sofia went to college, Andy enjoyed making videos of The Cuteness talking to her using the My Talking Pet app. Even the mention of it makes Sofia roll her eyes. "He thinks that's sooo funny," she says with total annoyance.

Andy loved to complain about the dog and his many transgressions. The Cuteness proved unable to learn anything, he seemed to have some mild OCD tendencies, and when left to choose between

Andy and Tommasina, he always chose the latter. Still, when The Cuteness had a near-death experience a year earlier, Andy drove him to Five Guys, letting him ride in the front seat of his Volkswagen convertible, and got him a burger. Andy thought The Cuteness was checking out and deserved a princely last meal. But the dog rallied.

This fall, though, The Cuteness stops eating and seems out of sorts. The day Andy gets word that the dog has cancerous tumors throughout his body, he breaks down in the lab, holding a counter for ballast. The kids are gone and he cries real tears and holds the back of his hand against his mouth to muffle gasps of grief.

His veteran students ache when they hear the news, as Cuteness stories are among their favorites. "Oh, my god. You won't believe what my dog did last night . . ." has been a common refrain.

But now Andy says, his voice limp and his eyes still wet, "I mean, what am I going to complain about now?"

He has to compartmentalize his grief because the demands of the forty-eight kids are starting to ramp up. The typical rhythm of the school year is a slow build and then a flash flood.

Exacerbating the chaos this fall is that one of the infrared spectrometers in the lab is broken—and despite several passes, Andy has not been able to fix it. The IR, as it's called, measures the vibration of atoms and helps identify the chemical composition of substances. Among the instruments in Andy's lab, the IR is key. It's one of the instruments he used to work on and he teaches the kids how to use it to great advantage in their research. He's tried swapping out different parts and still no go.

"We've got to get this thing working," he says, with a look of doom on his face. "So many of the kids need it for their projects." He starts reeling off names and it becomes clear the nonfunctioning IR will really gum up the workflow.

He's got such an intriguing mix of kids this year. The nerd factor is relatively low, though even if it were high, that wouldn't bother Andy.

He loves a good, hardworking nerd and takes great pleasure in trying to help with denerdification, intervening in their love lives, their wardrobes, their social acumen. Tommasina finds it amusing and admirable that her husband gets so invested in his kids, that he can't help himself. As a career chemist who runs experiments all day, she openly says that between the two of them, "He's the people person."

As always, there are some kids who require more heavy lifting than others, who need a steady stream of reassurances and support from him, who haven't yet gotten their research sea legs.

But he's also got some standouts, the ones who live in their own stratosphere. At the top of that list is Shobhi Sundaram, the self-taught ace coder who loves anything having to do with artificial intelligence. Shobhi has large, pensive eyes and always has her shoulder-length black hair tucked behind her ear on one side. She's recovering from a tennis injury to her wrist, so she's wearing a brace and using a special ergonomic keyboard. A self-possessed kid who talks at the tremendous speed that her brain works, Shobhi is developing a machine-learning model whose algorithm detects premalignant pancreatic cancer from blood samples.

There is no routine screening for pancreatic cancer, and by the time people are symptomatic, they usually have advanced disease. Ninety-one percent of pancreatic cancer patients will die within five years of diagnosis, with 74 percent dying within *one* year, according to the Hirshberg Foundation for Pancreatic Cancer Research. Shobhi thinks she can create an algorithm that will look at someone's blood and the proteins therein and accurately predict their likelihood of developing pancreatic cancer, thereby allowing for earlier and possibly lifesaving treatment.

One day, the Shobhinator comes to the lab with some news. "I got an email back from someone at MD Anderson," she says. She's trying to persuade the academic medical center in Houston to give her anonymous patient blood sample data for pancreatic cancer that she can use for her project.

Andy knows that it would be an incredible rarity for a hospital to give data to a high school student—Danny was turned down several times last year when he tried. And so when MD Anderson declines to help Shobhi, she doesn't skip a beat. She discovers publicly available mice data; that is, mice that have been engineered to develop pancreatic cancer.

There are hundreds of proteins in these mice blood samples. The critical question is, Which of them are the key players in pancreatic cancer?

Shobhi builds an algorithm to train a computer to pinpoint the crucial proteins. She works in Python, a coding language she taught herself. She identifies flaws in similar models that lack the ability to best capture the key proteins and so develops her own method, a fusion of preexisting ones.

As a first step, she weeds out the proteins that seem to be insignificant. She does much of this work in her bedroom at her home in Riverside. It's a process that can be repetitive and tedious. And because it's all computer based, there's nothing in the way of hands-on experimentation. She's the only kid in the class whose hands will never touch a beaker, flask, or test tube. But the cerebral aspects to coding and computer languages, the labyrinthine thought processes, are what excite Shobhi.

Her parents, immigrants from India and England, both of Indian descent, work in finance. Chitra, Shobhi's mom, introduced Shobhi to a simple coding language called Scratch when Shobhi was ten. Chitra had dabbled in coding in college and she thought it might be fun for her kids to learn. She had no idea that Shobhi would propel herself forward as she has.

Now, at sixteen years old, Shobhi's one of the most confident kids in Andy's class. Besides tennis, she does debate. She stands out for her boldness and her utter lack of timidity about her work. "In eighth grade, I started teaching myself [to code]. I had to really commit. It actually came pretty easily. I have a pretty decent memory." One of the more frustrating and disheartening gender dynamics among the science research class is that despite clear intellectual equality between girls and boys, the boys bear a confidence and forthrightness that just isn't there in most of the girls. Despite the girls being total powerhouses, most are reluctant to showcase themselves or their accomplishments in any assertive way. They need to be drawn out and almost prodded to take the credit they so deserve.

Not so for Shobhi. She's a star and she's not afraid to sparkle.

Which was why last year's CSEF snub hit her especially hard. During the first day, the judges examine and assess every student's project poster, the lengthy board that details all their work. They place finalist ribbons on those that advance. The next day, the finalists are invited back to deliver the pressure-cooker oral presentations.

Shobhi had developed an algorithm to determine which cancer drugs are most effective for breast cancer tumors. It was the only project in the category of computational biology. And of the two-dozen-plus kids from Andy's class who submitted projects, only she and one other kid were not named finalists.

Andy was floored and so were the Sundarams. "I don't think the judges got it," Andy says. "I don't think they had someone there with the background to understand what she did."

Says Chitra, with understandable exasperation, "They left a note on the project saying where's the experiment?"

Well, that's just it. Coding in this context doesn't render a physical product. Shobhi wrote a program that indicated with 87 percent accuracy which drugs were effective for various breast tumors. One needed to simply read how she developed the program and its efficacy—how she trained her computer model to answer these questions.

"That was the first time someone has questioned her work product. She'd come into it feeling very good," says Chitra. "The most difficult part of it was seeing her cry. I have never seen her cry like that. It was also the expectation, she obviously had such a certainty."

Chitra immediately called Andy. The two had a history. Long before Shobhi was in high school, Chitra had read about Andy's science research program in the local papers, which often chronicle the victories of his kids. When she and her husband, Vijay, decided to move from Stamford to Greenwich, she made an appointment with Andy to get his advice on which middle school he thought was best. She said to him, "This is the kind of course I'd be so thrilled if my daughter was in."

The Sundarams were all delighted when Andy accepted Shobhi, and for Andy, it was an easy call. Shobhi was an evident self-starter, a plug 'n play research kid.

When Andy got Chitra's call on the day Shobhi was not named a finalist, depriving her of a chance to deliver an oral presentation, he was still at the fair, trying to keep track of his students. In full protective mode, Chitra wanted to write a letter to the judges about their decision not to advance Shobhi's project.

Andy told her, "Look, if you do that, you're not only going to ruin things for your kid, you're going to ruin it for all the other kids in this program." Chitra backed down.

She certainly wasn't the first aggrieved parent Andy has ever had to deal with. Whenever one of these Molotov cocktails lands in his lap, Andy moves to defuse it. He listens, he empathizes, he acknowledges the slight. But he refuses to indulge or encourage the idea of taking things up with the fair officials. He's a big believer that lighting a forest fire for your kid isn't remotely helpful—that at the very least you become a pest, and at worst you look like a lunatic. He's seen how high the emotions run and how blinded people get when they think their kid's face has been rubbed in mud.

Plus, here's the most compelling reason parents in these situations need to stand down: all the ire in the world isn't going to change the outcomes. If the judges wavered on any of their decisions based on pressure from whiny parents, the whole competition system would implode.

In the grander scheme of things, with the passage of time, Chitra says she thinks the experience was a valuable one for Shobhi, that life is full of unfairness, and it's better to learn firsthand. She also thinks it's made Shobhi that much more careful.

"It's going to happen again," says Chitra, "and she has drawn a lot from it."

Another intriguing kid this year is Rahul Subramaniam, a sophomore who last year was the rare freshman in science research. Rahul is a bit gangly, with long arms and legs, a short torso, a winning smile that reveals mega-orthodontics, and a mop of black hair that either has a mind of its own or has never seen a comb. He has a thick lisp that somehow is charming. He wears funny T-shirts, like the one with some of the elements on it that says, "I wear this shirt periodically."

When Rahul arrives in the lab in the fall, he declares, "I think I want to do something on Zika virus." Andy has observed that on the heels of Olivia winning the Google Science Fair for her Ebola test, there are several kids who are drawn to newsworthy projects, as well as ones focusing on detection/diagnosis. He's not a huge proponent of this path because he wants the kids to be driven by their true interests and not by what they perceive to be an easy road to victory with a trendy project.

Zika virus is the mosquito-borne illness that can cause microcephaly in the babies of pregnant women who become infected. Microcephaly is a severe disorder characterized by brain damage and small head size. When Zika hit Brazil in 2015, there was an onslaught of haunting images of babies with microcephaly. Rahul says that's what drew him to Zika—and wanting to do something about it.

Rahul's father, Prem Subramaniam, Ph.D., is a microbiologist at Columbia who drives from the family's condo in Cos Cob (one of the towns in the greater Greenwich area) to upper Manhattan each day to work in a lab.

Prem is an occasional presence in Andy's lab, too, sometimes visiting the class to oversee Rahul's work. He also makes himself available to kids who may have questions about their projects, particularly those with a foundation in biology, such as Collin Marino, who's trying to get RNA to kill unwanted cells.

Rahul swats around a few ideas, ultimately deciding that he wants to design a cheaper, faster way to see if mosquitoes are carrying the Zika virus. The current method involves trapping batches of mosquitoes in areas believed to be affected, then sending them to a lab, where they're ground up and tested for the presence of the virus. It can take two to three weeks to get an answer, which is alarming considering how rapidly the virus can spread.

Rahul wants to do a real-time test, one that hinges on mosquito spit. Here's what he's thinking: If Zika can be detected in mosquito saliva (unknown), could he develop a solution that changes color if an infected mosquito spits into it? (Really unknown.) One hurdle he's cleared: the literature says that mosquitoes do, in fact, spit. Well, phew to that.

As much as Andy kind of groans at the disease du jour projects, he also knows that, in Andy-speak, "they get a lot of juice" at the fairs. So he doesn't deter Rahul. On the contrary, he's intrigued to see what this kid will produce.

15

William

October 11 is one of those autumn days where every color sparkles in a gem tone—the sky is blue topaz, the grass is emerald. In an instant, life has the optimism of a detergent commercial. It's the weather of wonderful moods.

It's Greenwich High's cross-country team's final home meet and it's being held at Tod's Point, a park and nature preserve along the Long Island Sound. It's the kind of place where you might propose to someone or stage the family photo for your holiday card. The light cuts in and out of the trees on a forested landscape and then breaks open onto the calm waterfront of the Sound. There isn't a more scenic place to decompress, revel in nature, or go for a run, even if it's among hundreds of other sweaty teenagers.

Several of Andy's kids run cross-country and it just so happens that two of them are among the school's best—nationally ranked junior Emily Philippides and sophomore Alex Kosyakov. Verna Yin, William's sister, is also a strong member of the team. Verna's so petite and doelike, with pencil-thin limbs, it's astonishing to see her move with such alacrity. She's light on her feet but has a force and purpose that belie her whispery nature. Takema Kajita, a stoic senior who joined the class the previous year, is also on the team.

And in a most surprising turn, so is William. Surprising simply because of William's natural glacial pace. "Like watching a turtle crawl out of a house fire," says Andy, shaking his head. When you hear about William Yin, academic bruiser, scientific genius, competitive pianist, fierce debater who can take you down in English *or* Mandarin, you might imagine someone with a commanding physical presence who barnstorms through life, a guy with a heavy stomp who eats his food in big, fee-fi-fo-fum bites.

But in fact, if you'd seen him just six months ago, in the spring of his junior year, you might have actually lost your own footing at first glimpse, as his stature and appearance stood in such sharp contrast to his near-mythical reputation. He had zero teenaged physical swagger and hadn't fully molted out of childhood. He had a build that barely made it to medium and seeing him stand in his living room wearing an oversized T-shirt, oblivious to the spiky cowlicks on the back of his head, it would be hard not to think, *This is William Yin?* You could imagine him drinking chocolate milk at his kitchen table and forgetting to wipe away his liquid mustache.

Now in the second month of his senior year, he cuts a different figure. His face has lost its childhood chub and he's filled out, looking more his age. His hair is still an afterthought, as he lets it grow over his ears and brow before allowing his father to shear it off. This often involves a weeks-long campaign on Mark's part. William still wears his blue wire-framed glasses, which he regularly props up by the bottom of the right lens, as if to steady his thoughts on the matter before him.

Unlike Verna, he doesn't consider himself to be competitive as a cross-country runner. "I do it to get some exercise, you know, stay in shape," he says. He mostly hangs out with Takema, or Tak, as the kids call him.

There are about five other schools competing at this last home meet, including the powerhouses of Wilton, Saint Joseph, and Danbury. The starting line is on a hilly section of the park shaded by tall pines. When the gun blasts, there's a blur of red, blue, yellow, and burgundy, as the pack takes off. William and Tak are near the back of the bunch, but you can see they're running their hearts out.

Alex Kosyakov, the first-generation Russian American sophomore science research student, wins the three-mile race with a time of 16:16. About twenty-four minutes after the starting gun sounds, William comes running toward the finish line. The field had scattered, and even though he's coming in near the back, just as he started, he's giving it a good, final push, speeding up and not slowing as he makes his way over the line. He has strong form, back straight, knees coming up to his waist, and as he finishes, there are shouts of "William!" ringing through the air.

Next up are the girls. Verna and Emily finish in the first third of the pack.

After both races are done, the GHS parent organization lays out a spread of hot dogs, cheesy baked ziti, potato salad, and cookies. The kids stand around devouring the food, immediately replenishing the calories they just burned.

And since it's the final home meet of the season, there's a tradition where the seniors do a mad dash into the Long Island Sound. There's a crackling fall chill in the air, making the ritual that much more daring. The boys make their way to the beach and start peeling off their running jerseys.

William and Tak stay behind, looking somewhat despondently at the rippling low tide. Like puppies who don't want to go out in the rain, they just hang back watching their teammates eagerly strip down to their shorts.

Tak is slightly more willing to partake in the festivities and slowly he and William start lumbering toward the group. Tak suddenly panics about his glasses. Who's going to watch his glasses?

"Excuse me, can I just put my glasses here? Excuse me, my glasses, um, if you could just—" No one's listening to Tak amid the chaos.

He and William are the very last senior boys to enter the frigid

Sound. Unlike their teammates who charge from the shore into the water, screaming and splashing and whooping, William and Tak slink into the frenzy, like two old retired men entering a wading pool (or two turtles crawling out of a house fire), their fingers skimming the surface. They don't yell or try to whip water onto each other. They actively move away from such antics. But they're there. They did it. With their lips quivering, limbs shivering, and not a small amount of concern about Tak's eyeglasses, they mark the last home sporting event of their high school careers in the icy blue water.

For William, he decided to take the (not quite) plunge the way he makes nearly every single choice before him: at the very last second possible.

By this time, William has made one decision about a new science project: he won't decide whether he's going to pursue one until after he's completed his early action application to Harvard. Whenever he's asked whether he intends to launch something in Mr. B.'s class, he says, "I really want to . . ." but then his voice trails off into oblivion and it all feels very wobbly.

He's also working overtime trying to sand over the deep grooves of procrastination in his brain.

"I'm really bad," he confesses. "It's gotten to a point where I've got to stop." Like many procrastinators, though, William claims that his habit fuels a kind of high performance adrenaline—and he believes he does some of his best work while up against the clock.

His early application to Harvard is the seemingly perfect occasion for William to reform his ways. He gets a jump start on the essay, taking his first crack several weeks before the November 1 deadline. He makes regular hour-long appointments with Ms. Patti to go over it letter by letter.

He initially drafts an essay about teaching himself to code. But the essay doesn't feel quite right to him, not quite revelatory.

So he scraps it. Since he's submitting both of his science projects

as part of his application, the cancer drug delivery/radiation en-
hancer nanoparticle and the atherosclerosis detection sticker, he decides to devote his essay to the other thing he prizes but is remarkably
modest about: his piano playing.

He started taking lessons at age six, noting in conversation, "It's a
typical Asian thing." Like many things, the piano came easily to him.

But what drew him in was not the technical mastery needed to
produce luscious, transporting music. "It was the emotional aspect
to it that really hooked me. I was drawn to the raw emotional aspect
to it," he says. "A lot of times, I make technical mistakes, so when I
was able to link the movement to actual emotional experiences, that
was when I started really enjoying it and that's what kept me going."

William went out onto the competitive circuit and started winning contests by playing Debussy, Chopin, Bach, Beethoven, Haydn,
Mozart, and Copland. He played as a Grand Prix winner at Carnegie
Hall, both his sophomore and junior years. And he and Verna, also a
gifted pianist, gave several concerts in Portugal in 2015.

But early on as a player, he came to understand what lit him from
within as a pianist, and it wasn't classical music.

It was jazz.

William loved the freedom, the incongruity of it all. He loved the
ability to allow his fingers and mind to drift to new, unknown places
and then bring them back home. Even with classical music, he says,
"I like the music that's most expressive and has the least structure to
it so I have the freedom to build emotional structures. Or build my
own story from the music."

Because above all, William prizes creative freedom "in any field of
anything." That is what both music and science give him. Profound,
unfettered freedom of thought.

And it is the open waters of his mind, the exploration and the lack
of rigidity, that feed both his science and his music. The best scientists, like the best musicians, are not merely those who are automaton-
like in their thinking and execution, who stay closely within
well-established bounds. No, the best scientists free their minds of

convention, reject preconceived notions, and wander boldly to unknown places to find new, more novel ideas.

So William crafts his Harvard essay about one of his most personal
and private escapades, which is occasionally sneaking into the
school's new $29 million concert hall and playing the piano all alone.
He loves the illicit yet harmless aspect to it, the surge of triumph he
feels for merely seating himself at the bench. And of course there's
the delight of playing the old Steinway and hearing his music fill the
belly of an auditorium, one big aural feast.

He writes:

I set my hands on the piano and cautiously press a single key. The note
emerges, unobfuscated. Its sharp clarity is a note of confirmation. All
of a sudden, my pent-up emotions burst forth upon the piano in a
flurry of melodies and harmonies. Glee and excitement spring out as
rapid dances of notes in the upper register; anger and frustration surface as booming, dissonant chords in the lower range. As my fingers
fly from one end of the piano to the other, the auditorium is filled with
a cataclysm of emotion as if from Pandora's box.

And yet, in my mind, there is no score, no notation on a page, no
musical schedule to follow. Instead, the music is completely impro-
vised, drawn from within me.

At the end of the day, I have formed nothing more than an ephem-
eral medley of notes reminiscent of a Jackson Pollock painting. As I lift
my fingers from the keys, I am once again aware of the heavy silence
of the empty school auditorium. There is no audience, no orchestra,
no one except for myself, sitting upon the darkened stage. The piece I
have just crafted and performed has receded into the stillness, never
to be played again.

The next day will be like any other. I will once again be submerged
in a scheduled, rhythmic storm of events and activities, zipping from
one place to the next with an exactness fine-tuned through many

years of experience. For many hours in a row, there will be no rest, no break from the ceaseless tempo of the metronome beat. And yet at the end of the day, I will once again find myself in that silent auditorium . . . Because, just as in music, the moments of silence are just as important as the music itself.

To understand the current of aspiration, hope, and pursuit that runs through the Yin family is to understand the childhood of its patriarch, Mark Yin.

Mark grew up in the densely populated Zhabei section of Shanghai, the second son of semiconductor factory workers. Mark says the neighborhood was "not desirable," and if they spoke too loudly within their two-story apartment, their neighbors could hear them through the shoddy walls. The family lived without heat, airconditioning, or a bathroom. They had running water, but to use the toilet or shower, they went to a communal facility that was a two-minute walk down an unpaved street. It's impossible to say that Mark Yin is currently the sole Greenwich resident who spent the first eighteen years of his life without a toilet and shower in his home, but if there are others, they are few.

In Mark's graduating high school class of 320, just four people went on to college. So rare was higher education for people from Zhabei, Mark and his three friends who went to university were known in the neighborhood as "the college students" and were viewed with great reverence.

Mark's family supported his academic endeavors but had no reference points for his achievements. He went to the prestigious Fudan University and then got tapped to come to the United States to do a Ph.D. at City College in Harlem.

He arrived on January 18, 1992, and immediately borrowed three hundred dollars in cash from his advisor so he could begin to establish a structure to support basic needs. He'd never traveled outside of China before, and even though he'd grown up in the large metropolis of Shanghai, New York City was so completely foreign in every way—its diversity, its seemingly natural state of constant improvisation, its pluck and vigor.

He came alone but sought to remedy that. Through a series of friends, he met a couple, Chinese medicine practitioners with a business in New York City's Chinatown. Word was the couple had a single daughter.

The first thing the Chinese gentleman did upon meeting Mark was take his blood pressure and heart rate, to make sure he was healthy. Once Mark made it past the initial screening, what followed has to go down in the annals of courtship as one of the most patient pursuits of all time.

Wendy was, in fact, single. However, there was one minor catch: she was living in China. In order to earn the right to an introduction, Mark embarked on a long vetting process by her parents. The decent blood pressure wasn't enough. He spent a good deal of time getting to know them, even vacationing with them in Maine. It was one long, drawn-out audition for a part it was impossible to know he definitely wanted, since he had yet to lay eyes on Wendy. Since this was the mid-1990s, before regular email or Skype, the two wrote letters and spoke on the phone. At one point, Mark got worn down and wanted to bail, but Wendy's parents encouraged him to persist.

About a year after he'd submitted to the vitals check, Mark traveled with Wendy's parents to China to finally meet their daughter in person. And despite their unorthodox courtship that spanned the globe, they both knew they'd found the one. A month later they were married. Mark brought his bride to the States and three years later, on May 30, 1999, William was born in Princeton, New Jersey, where the Yins were living for Mark's job at a pharma company.

Verna arrived two years later, on May 18. The growing family eventually moved to Greenwich, attracted by the stellar school system and proximity to the city, where Mark commuted for twelve years to Albert Einstein College of Medicine, doing medical research.

In 2015, Mark called a family meeting. He told William and Verna he wanted to leave Einstein. With just a few years left before both kids were in college, he wanted to be at home. He asked the kids how they felt and they expressed their support. To make ends meet, Mark started a tutoring business, mostly in math, and launched a cultural consultancy that helps facilitate Chinese sports teams when they compete abroad. His company arranged for the Chinese national swim team to train for a couple of weeks in Connecticut ahead of the 2016 Summer Olympics and Mark then accompanied the team to Rio de Janeiro. Wendy works part-time as a bank teller in town.

As is the case with any first-generation immigrant kids, William and Verna will never travel as far as their parents to arrive wherever it is they're headed. They will never know what it's like to walk on a dirt road to take a shower, or to land in a foreign country in winter clutching a promise of better things, but with no money or guideposts. They will not stare across that same mental and emotional canyon formed when a child soars far beyond their parents' lot in life.

But palpable in any conversation with Mark and Wendy is their yearning for their children to have an easier and less rocky climb, to know a happiness that is earned but less hard-won than their own. Mark doesn't regale his kids with tales of how completely unlikely and remarkable it is that he earned a Ph.D. after growing up in a place where owning a backpack was a sign of status. But you need only look at Mark and Wendy's faces and step into their home to see up close the love, the pride, the sacrifice.

And the dreams.

16

Romano

Part of the homecoming tradition at Greenwich High School is a parade down Greenwich Avenue. As people walk their dogs up the Ave., doing the first world balancing act of latte, leash, and iPhone, the high school pep band belts out brassy tunes against the snap of snare drums while a line of floats from a variety of school clubs and sports snakes it way southward from the Post Road. Candy and lollipops rain from the air as the students dance and toss sweets to the parade watchers who line the street, putting their morning errands on pause to witness some Cardinals spirit. The kids are earnestly enthusiastic, waving and dancing on top of the floats, absent of any teenaged smugness, embarrassment, or inhibition.

Toward the end of the floats are a few candy-apple-red convertible classic cars chauffeuring the homecoming court, school headmaster Dr. Winters, and the football coaches. Everyone's wearing red, the school color, waving and smiling—basking in their royal moment.

Romano is sitting beside Caitlin atop the backseat of the car, blowing kisses to the crowd. There's no apparent chill between them, despite the fact that he still plans to end things with her. He's wearing a smart red plaid shirt beneath a red sweater—an almost sock hop, retro look.

It's October 15, a sun-drenched Saturday. The once jade leaves have crisped to the full autumn palette, and the whole scene is a picture of Rockwellian Americana. If aliens wanted to understand what homecoming is, boom, this is all they'd need to see. The kids at Greenwich High often lament that the place is lacking in school spirit, that the sheer size eradicates the ability to rally the masses to any kind of rah-rah. But today, it's all on display.

The night before, Romano and his posse tore it up at the homecoming dance. One of his best friends, Greg Goldstein, junior class president, warned everyone in the student government that attendance was mandatory. He was sick of the default chain reaction of people not going unless the right people were there, which led to lackluster attendance for fear of being at a lame event. Show up, said Greg. It is what you make it.

Romano got there early and was instantly deflated at the state of the party. Gaggles of kids were crowded around the edges of the student center in awkward bunches. He scanned the room and immediately spotted Matt Armstrong, a kid he's known since middle school, dancing out on the floor by himself, seemingly happy. Romano was chosen to give the middle school graduation speech and he invited Matt to deliver it with him.

Matt has Down syndrome. He runs cross-country and participates in a variety of school activities. There's a school support committee for Matt that helps chart his PATH (Planning Alternative Tomorrows with Hope). Spearheaded by Matt's mom, the committee includes peers, teachers, and administrators. Romano is a member of Matt's committee.

When he sees Matt dancing on his own, Romano thinks it's bullshit. No one should dance by himself at a school dance. He makes his way over to Matt and starts throwing down. Matt's happy to see Romano and soon the two of them are shaking their asses to the music.

It takes less than a song for people to start moving in to join the two of them. The crowd forms a circle around Matt and Romano and as kids shed some of their self-consciousness, some of them move into the middle of the circle.

Greg jumps in and he and Romano do a bunch of their signature moves. Romano tosses an invisible fishing line and Greg plays the fish who gets reeled in. Greg pushes an invisible shopping cart down an aisle, chucking food in his basket. Romano pretends to shake the milk. All this while the DJ is spinning new and classic dance tunes, everything from Drake to the Spice Girls.

"I loved the dance," says Romano, the Monday after homecoming. "I had so much fun, so much fun."

He enjoys the rest of the weekend too, relishing his role as a kind of homecoming Zelig. There he is at the pep rally, the dance, the parade, and now at the actual football game. If he has any tinges of regret over not being on the field, none are apparent. He's sitting in the stands with the pep band, wearing their signature gray sweatshirt over his red sweater. Romano plays cymbals, his gesture of school spirit. William plays mellophone in the pep band and Alex Araki, a fellow science research student, plays trumpet. Whenever the band fires up, Romano cheerfully clashes his cymbals.

The game has a decent turnout, with the stands nearly full. At half-time, an emcee strolls out to crown the homecoming court. Romano removes his band hoodie, yanks down his sweater, and heads to the field. The court is populated with a few underclassmen princesses and princes, but just one king and queen, seniors. The emcee announces each title and calls the name of the coronated, each member of the court taking his or her place in a line across the field.

"And junior homecoming prince . . . Romano Orlando!" He steps forward, receives a red carnation boutonniere, and has his picture taken.

Afterward, he and Caitlin pose for more photos on the sidelines. Kim and Rome are at Sophia's parents' weekend at the University of Arizona, so both of Romano's grandmothers and assorted other Orlandos are at the game, snapping pics. He seems genuinely happy to be here; the knowledge of what lies ahead with Caitlin doesn't appear to taint the festivities.

Romano isn't much in the way of a contributor to social media. It mostly doesn't occur to him to document and share his life the way many of his peers do. But it's not every day you're named homecoming prince of your class and get chauffeured down the Ave. in a classic car, so when it's all said and done, Romano posts just one photo to Instagram of himself from homecoming weekend. It's of him leaning down and getting a giant smooch from . . . Nanni, his grandmother.

She's always loving on him hard, kissing him as if every time could be the last time. With her limited English, each kiss seems to be a little poem of things she feels but cannot say. While she does this, Romano emits sounds that are a combination of giggle and happy guttural groan, like a puppy getting nuzzled. Anyone who knows the sacrifices Nanni's made for her family could get teary watching this, because *this*, right here, right now, the kissing and loving of her youngest grandchild, is what has made it all worth it.

Love is many things, but in the form of an Italian grandma, it is at its most fierce and visceral. And caloric. God, Romano loves Nanni—and her cooking. On Sunday night, when the homecoming hoopla has come to a close, Romano heads over to Nanni's for dessert. He has this to say about his cardigan-wearing grandma from the Old Country: "We're beyond tight, she's my woman . . . we chill hard, we roll deep."

When Romano walks back into the lab on Monday, he's jubilantly exhausted. Being popular is hard work. As for sharing in all the homecoming glory with Caitlin, he says, "What this weekend did, it sort of reminded me why I liked her in the first place."

He sits down at his usual spot in the lab, center back table, and gathers his reading materials. For his liquid bandage, he's spending class reading everything he can about the amino acid L-DOPA and trying to design his project. L-DOPA comes in powder form. He's asked Andy to order it. He'll need to add it to a substance that will harden. The L-DOPA, he hopes, will bring the adhesive factor—it'll be the component that makes his bandage stick. After reading and speaking to Bruce Lee and Mr. B., he's thinking he'd like to infuse L-DOPA into gelatin, since the latter cures (firms up) quickly, which his liquid bandage will need to do to be effective. And he still thinks adding an antibiotic and enhancing the healing properties will make his project next level. But he's not committed just yet.

What he is committed to is ending things with Caitlin. As much fun as homecoming was, it wasn't enough to steer things back onto the relationship highway. Since he's looked for but hasn't seen any signs of Caitlin sensing the end is nigh, he decides not to instantly shatter the homecoming euphoria with the breakup sledgehammer.

"It's going to be this weekend," he declares with a tone of conviction. In terms of how exactly it's going to happen, he's fretting about the details.

"Well, I think her mom's going to drop her off at my house. But then, we, like, hang out, I tell her the news, and then what happens? Is her mom going to have to come all the way back to my house to get her? She lives on the other side of town. That's awkward. I mean, am I going to have to get her an Uber?"

Oh, the perils of the contemporary teenage suburban breakup. When you think about it, having your newly ex-boyfriend Uber you home maybe doesn't seem so bad? Better than having to call your mom for a pickup shortly after drop-off?

When the weekend arrives, the plans change. Here's how it's going down: Romano's getting dropped off at Caitlin's house. Her parents won't be home. He's not going to stay long or beat around the bush. He's going to lay it out and then leave. No sense in lingering. He's positioned Greg on call for pickup. He'll text Greg when he arrives at Caitlin's and then Greg will know to pull up about twenty minutes to a half hour later. He'll roll up, Romano will walk out, and they're off. Greg's got his back, he's reliable. He *is* junior class president, after all.

And so on Saturday night, Romano shows up at Caitlin's. He's not trembling, but he's definitely nervous.

Five minutes after he arrives, he gets a text from Greg saying he's here. Here being . . . Caitlin's driveway. What the fuck? Why is he so early? Now what? Is Romano supposed to make it snappy? Goddammit, Greg. The overeager getaway car is not making this easier. Now Romano's a bit distracted thinking of Greg sitting in the driveway— and hoping Caitlin doesn't happen to notice. (Fortunately, she has a very long driveway and her house is set back from the street.)

First, the imminently former couple make small talk. "We talked about her friends and the election and the debate. I could tell I was off," Romano says. "She was sitting down and I said, Come sit on the couch." And then, partly because he hadn't scripted his breakup speech, partly due to how legit important his liquid bandage had become to him . . . he blamed science research.

"I actually kind of threw Mr. B. under the bus," Romano says, tittering just thinking about it. "I told her that now that I'm in science research, I'm going to have to be staying after school a lot. I just got my molecule and it's heating up."

I just got my molecule and it's heating up. Said no one ever in the history of breakups—until this very moment on Saturday, October 22, 2016.

He continues. "I just don't have the time to be the boyfriend I could be to you . . ." Back to the playbook. Phew. Because no girl wants to be dumped FOR A MOLECULE.

"I can't be that person now. I'm either exhausted or busy. I can't even text . . . We hang out once every three weeks. And I don't get to hang out with the boys."

Caitlin starts to cry, Romano gets weepy himself. He's surprised at his tears, but she was his first girlfriend, after all, and he fought hard to defend and maintain the relationship. Of course, there's relief, too.

He had the hard talk. But he's genuinely sad and has that post-split hollowness, that cavern on your insides that was once comfortably filled by the other, now heartbroken, person.

Says Romano, his voice sad but steadfastly resigned, "In two minutes, the eight months were all over. How absolute and how fast."

17

Andy

On October 25, Andy gets an intriguing phone call from Bob Wisner, the director of the Connecticut Science and Engineering Fair, the state's oldest science fair, which is turning sixty-nine this year.

Bob is a CSEF alumnus whose 1958 and 1959 wins sent him on to the National Science Fair, which would become Intel ISEF. For a working-class kid from Hartford, Connecticut, the 1958 national fair gave him his first plane ride—to Flint, Michigan—where the event was held that year.

Bob went on to have a successful thirty-eight-year career as an engineer with what became United Technologies, specializing in high-energy lasers, optics, and high-powered radio frequency sources used for things like simulating nuclear-powered rockets. He says his proudest accomplishment was leading a team that developed a gait analysis system that helped evaluate kids with cerebral palsy.

Bob is an example (there are many) of people who felt their life trajectory was changed and shaped early on by their successful run on the science fair circuit—that the exposure to independent research and the joy of victory pointed them to a gratifying life of science.

"It gave me my career. Period," Bob says. Now seventy-five, Bob has been involved with CSEF since the early seventies and fair director—an entirely volunteer gig—since 1989. The allure for him is simple. "It's the fact that I can see the next generation of kids," he says. "We're providing opportunities, recognition, and in some cases mentoring for the next generation of scientists and engineers."

Because of his stature and longevity with CSEF, when Bob's name pops up on Andy's phone, Andy picks up without hesitation.

"So there's this other fair, the CT STEM Fair, that's been around for years," Andy says, a bit of curiosity in his voice, the day after speaking to Bob. "I've never brought my kids there because it's a month earlier than CSEF, so that means you have to get everything done a month sooner. Which, with fifty kids, is really tough.

"I've actually been a judge at CT STEM and I took Sofia there one year to compete. But it just never had the juice of CSEF. The prizes aren't that big, and again, being it's in February, I said, forget it, not worth our time.

"So, now Bob calls me and he says that CT STEM got ISEF affiliation this year. They have three spots for kids to go to Intel."

Andy is sitting at one of the lab's low tables, laptop open. You can see his mind is whirring like a pinwheel in a hurricane at this news. He's been hoping and waiting for this call for two years, when he first got wind that CT STEM might be on its way to gaining ISEF affiliation.

For as long as anyone can remember, CSEF has been the only way for Greenwich kids to win a berth at Intel ISEF, which is an unrivaled extravaganza, the Olympics of the science fair universe, drawing kids from seventy-five countries, regions, and territories. CSEF awards seven spots to its top kids, including one that must go to a student who attends an urban school. State and local fairs around the country have to apply for Intel ISEF affiliation, which is still the main way for kids to get one of the 1,800 Intel ISEF slots.

Every year, Andy sends his students to several local fairs and competitions, but a great deal of emphasis has always been placed on CSEF. He works up until the very last second getting his kids ready, running around like a madman until fair day, so the thought of squeezing himself into a vise grip a month earlier has never held any appeal. But these three Intel ISEF spots . . . this means three more golden tickets for his kids and another opportunity for them to dominate.

Andy wants in.

Over the past ten years, Andy has sent at least one and as many as seven kids to ISEF each year, for a total of thirty. Cumulatively, he's sent more kids to ISEF than any other teacher in Connecticut state history. No one tracks such a distinction on the national level, but considering this is only his eleventh year teaching research, he's likely part of a small, elite group of individual teachers who have sent so many kids to ISEF.

Which has been a source of awe and ire in the Connecticut state science fair community. There's no shortage of grumbling that it's unfair that so many kids from one school should be allowed to take so many spots. And the fact that the school is in wealthy Greenwich doesn't exactly engender goodwill. Plus, sometimes the Greenwich set doesn't do itself any favors. One year at Intel, one of Andy's kids ordered himself seventy-five-dollar king crab for dinner—on CSEF's dime. In the moment, the chaperone said it was fine, but Andy says that after the fact, she "ripped me a new a-hole" in an email. "Never been the same since."

Nearly all of the griping makes its way to Sandra Müller, president of CSEF. A retired director of nursing who could easily be cast as Mrs. Claus in any Christmas movie, Sandy first got drawn into the science fair orbit three decades ago when one of her own sons competed. She and her husband, Wynn, joined the CSEF volunteer corps in 1995, performing a variety of functions, including chaperoning kids to ISEF, which currently rotates annually between Los Angeles, Pittsburgh, and Phoenix.

"When the boys were younger, we were involved with Cub Scouts and Little League. It was not as satisfying as watching these kids create and learn and explore the science," says Sandy, sitting in a reclining armchair in the Cromwell, Connecticut, home where she and Wynn have lived for some forty-five years. The house is a virtual museum of the 1960s, right down to the original brown enamel appliances. Their waste-not, want-not thriftiness has been a boon to CSEF. Sandy doesn't let a dime go wasted and knows how to stretch the fair's annual budget of roughly $185,000, which does not include the $170,000 in prizes and scholarships it awards.

Sandy takes pride in the fact that CSEF has no paid staff, save for an occasional photographer on fair day, which even extends to her, the fair president. Her duties are year-round, and given how involved they are, it's extraordinary and inspiring that she's been doing this for just about twenty-five years.

She and her team must find corporate sponsors to help cover costs and prizes, secure a venue, solicit volunteer judges from the community, and review all applications, a painstaking process to insure that the kids' work is safe and ethical.

Over the years, there have been a number of attention-grabbing project applications. Sandy once reviewed a paper where "it appeared that they had buried bodies. One decayed faster if they wore polyester. They were taking calves' livers and made clothes for them—but nowhere in this paper did it say that it was a SIMULATION."

One year, a kid was building a rocket whose materials reminded the person reviewing the project of those used in the Oklahoma City bombing. His entry was denied. Another time, a different kid's rocket project triggered an alert to contact Homeland Security due to the websites he was searching for his work.

"Had to deny the project," says Sandy with a shrug.

Sandy and her team also must sometimes perform the unpleasant work of auditing projects they suspect have extreme input or oversight from mom and dad.

"You can tell right away. You call on the phone and you say mom or dad are free to be on the line and when mom or dad monopolize the conversation and the student knows nothing and you say, 'Where did you get your specimens for milfoil?' and they don't have any idea and the dad says, 'Candlewood Lake . . .'

"If a judge doesn't pick up on that from an abstract [scientific brief], most of that shakes out in the interview process."

And then there are attempts at sabotage—anonymous tipsters accusing participants of fraud, such as plagiarism or outright fabrication of results. Each of these cases is taken seriously and investigated in order to insure the fair is . . . fair.

"Well, we haven't involved the FBI," quips Sandy, in a tone that suggests she might not be above doing so.

The final kick in the teeth is what Sandy calls "the nastygrams," missives from parents whose kids don't advance at the fair. Here lies an interesting window into contemporary competition/parenting mentality. While most parents of said kids do not devolve into something resembling a cornered, fanged desert animal, some, in fact, do. These men and women are the science fair equivalents of the parents who scream at the Little League coach or ref and get tossed from the sidelines. The science fair version is ever so slightly more genteel, usually choosing to voice their rage via email. (Andy gets a hybrid—emails and phone calls.)

The drama around Andy's kids is largely different, primarily because Andy's approach reduces much of the potential for fraud.

Most schools farm their science-research students out to universities or professional labs so that each works with a mentor, a scientist with expertise in the kids' area of interest. The work is largely done off-site where the kids have access to professional-grade equipment and materials. They're supervised by research assistants, postdoctoral lab staff, or even the lead scientists.

"The bad feature with those programs," says Sandy, "is you have to be very careful that the students are doing parts of it themselves. That it's not being done by lab assistants, unless it is a specific part that is dangerous. There are lots of chemicals that a student cannot touch. If a lab assistant is loading the gels in the slides, we don't want the students to just look into a microscope and say, yup, that's a cell."

Other schools often make science research a multiyear commitment. Initial years are spent analyzing papers, learning lab skills and the scientific method. A kid might not start to do hands-on project work until more than a year or two in.

Andy dispenses with both of these models. "The kids don't have the patience or attention span for that," says Andy. And neither does he. "The kids need to have ownership of their projects, to feel like they're really working on something and owning it."

From the day he arrived in the lab, drawing on the format and methods of Ray Hamilton, his predecessor, another above and beyond teacher, Andy set up his class so that he and the kids would be on a high-octane ride, completing projects in a single year. Because of Andy's side business repairing and calibrating various instruments, he's always got a bead on a machine that a professional lab no longer wants. For this reason, his lab has enough professional-grade equipment (minus the scanning electron microscope he covets) for the kids to do their work on-site.

Andy would have it no other way. To him, the whole reason he got into the teaching business was to work side by side with kids, to develop the relationships and let the science unfurl in all of its glorious unpredictability. He thrives off the connections and the adventure—for both, he's willing to go where few would and has put up with a good deal of shit in the process.

Actual shit. Andy has done everything from taking a kid to a sewage treatment facility to collect water samples from a football-field-sized pool of bubbling poop to visiting a sludge site in the Bronx's Hunts Point neighborhood that baked human excrement into fertilizer pellets. The latter was in one of Andy's first years teaching research and he earnestly wore a suit to the poop plant, wanting to look professional. The guy escorting Andy and his student took one look at him and said he was overdressed for the occasion.

The site actually dealt with fresh, steaming truckloads of human feces. Andy pointed to a guy in a full-body rubber suit hosing the poop detritus off the exterior of the truck and asked, "What does he make?" (Incomewise.)

To which the company guy said, "You won't believe what he makes. But they never last. They leave within a couple of months."

The stench ranks as the number one worst odor Andy's ever smelled. It left such a pungent scent on his suit, he tossed it the minute he got home. "I couldn't imagine handing it over to a cleaner," he says.

Notably, the New York Organic Fertilizer Company shuttered in 2010. A press release from the Bronx borough president's office acknowledges the nuisance of the "vulgar stench" and says, "Hunts Point residents are glad that, after years of complaining about foul, noxious odors, their long-time activism has resulted in the City refusing to renew its contract with NYOFCO, bringing us one step closer to removing NYOFCO and its disgusting emissions from their neighborhood entirely."

So mentoring all of his kids through their projects sometimes means embarking on occasional field trips to vile places. (When you've been to a sludge site, the state mosquito research center seems like a carnival.) But the alternative, the "farm out" model, as Andy sees it, is more like project management, a teacher keeping tabs on kids who are off working with professionals. Andy was once one of those professionals. This, he thinks, is the sine qua non of his value—both his background in industry and his willingness to devote copious amounts of time to his kids.

Occasionally, if something falls far outside his expertise, he'll enlist help. For instance, at this stage in life, he's not going to teach himself to code.

But he mostly takes the other route, which is good old-fashioned elbow grease. When a kid brings in an idea that's somewhat foreign to him, he'll dive into the literature and bring himself up to speed so he can backstop the work. It's not unusual for him to read a stack of published research papers on one topic to prepare for a kid's project.

Are there problems with Andy's approach? Yes, the main one being that he's a one-man band to forty-eight kids with demanding projects. In another school, his kids would likely have forty-eight individual outside mentors who would guide them and keep them on track. People have noted how hard and how much Andy works and regularly take his temperature to see if he's on his way to self-immolation. He's not infrequently asked if it's time to start looking for his replacement. But Andy likes the reputation he's earned, the cult of personality around Mr. B. Years back, an administrator suggested he join forces with another science teacher to co-teach research in order to ease his burden. Anyone who really knows Andy knows he wouldn't be comfortable with that. The research class is his kingdom, and not one he's looking to abdicate to or for anyone. He believes there are few teachers willing to put in the kind of hours he does—judging by the crew of clock punchers among his colleagues. And it's that time, plus his knowledge, that breeds a great deal of the success of his kids.

Says Sandy, "The Greenwich parents, while involved, don't always know what their kids are doing because these kids do their own work. Andy is their mom and dad; he spends an inordinate amount of time. Those students respect him, the parents respect him, and I know the administration there respects him. I truly think he gets acknowledged for what he does."

And here lies the most pressing issue for Andy. He's not so sure about that last part. Is he widely known as a superstar teacher? For sure. But the prior year, there was an incident that for him cast serious doubt on whether the powers that be actually value *him*, as opposed to the glory his program bestows on the school.

He went to an administrator and asked for a modest raise that would have amounted to about $4,000 a year.

The point, in Andy's mind, wasn't the money, which would hardly be life changing. Now in his twelfth year at Greenwich High, he earns close to $120,000 a year, which includes an extra $4,000 for heading up science research. Teachers who run clubs or coach sports get the additional money, too. Greenwich teachers are among the best compensated in the state of Connecticut.

What he was looking for was a symbol of respect and appreciation, a token that suggested that in the school's eyes, he was valued and worth it, and that in the scheme of things, $4,000 was nothing, considering both what he puts in and the results he gets.

Raises for public servants are not as simple as a yes or no. There are pay scales and procedures. In the end, the administrator told Andy to call the district's human resources person to inquire about the possibility. What Andy was hoping to hear was "Let me make a call for you and see. I'd love to help you out." He wanted a show of support, for someone to punctuate his worth by picking up the horn on his behalf.

He was indignant at the idea of calling HR himself. "If you think I'm going to grovel for four thousand dollars, forget it," he says, his irritation bubbling over. "That's not the point."

The whole thing doused some gasoline on his attitude and reinforced his suspicion that he was not fully appreciated. And that's really the heart of it. Andy will do the work, but he wants to be differentiated from the teachers who are in the parking lot before the 2:15 bell has finished sounding. He draws a line between himself and some colleagues who do the minimum, are shielded by tenure, and appear to be counting the years until they make the magical twenty, when they can retire with a handsome pension, provided they fulfill certain criteria.

Those who have speculated about Andy's potential for burnout aren't wrong to wonder. From November to May, he puts in sixty- and seventy-hour weeks, including nights and weekends. Even though the job is completely in line with his genetic predilection for hustling and inability to be idle, eleven years into research, there's a creeping but steady sedimentation of fatigue and a realization that not only is his setup unsustainable until retirement, it's probably not what he wants long-term anyway.

Which is why he's been going to night school to get his school administrator's certification. At the very least, he'd like some options to become a department head or, in theory, a principal or assistant principal. He'll likely be done with all the coursework and internship hours this year, and so slowly, he's starting to look beyond room 932.

18

Ethan

When Ethan Novek was three years old, he ran into his parents' bedroom in the wee hours of the night and woke them from their sleep to address a burning question on his mind: Were there 64 ounces in a half-gallon milk container or more than 64 because of the little triangle roof on top of the carton?

Stop and think for a second what you might do if your child did such a thing. Most people would have likely said, "I dunno, go back to bed." Some may have tacked on a conciliatory, "We'll figure it out tomorrow."

What happened next in the Novek household will make so many other subsequent things in Ethan's life seem logical.

Keith and Bonnie Novek listened to Ethan's question, confessed they had no idea whether the triangle roof on the top of a milk carton meant extra ounces, and said they'd go find out. So in the middle of the night, the Noveks trudged off to the kitchen, took out a fresh half gallon of milk, and poured it out into a measuring cup in several installments to count the ounces. Ethan got his answer and could go back to sleep, which meant so could his parents. A half gallon of milk contains precisely 64 ounces.

Ethan's curiosity didn't end with the milk carton. It was just beginning.

The first time Keith thought his second son might be unusually bright was when Ethan was five. Ethan, who his mother describes as a late reader (which really means that it was one of very few things he didn't do at an accelerated pace), started keeping a journal of scientific inventions. In it, he drew his ideas for things.

But as he got older, drawing blueprints for his ideas wasn't enough. He wanted to get his hands on materials and start experimenting and building. In fourth grade, he was home sick from school one day. He had been reading about electrolysis, the process that uses a direct electric current to produce a chemical reaction. Luckily, Ethan had plenty of copper wire lying around the house because wire was one of his obsessions. He spent a lot of time at the local True Value hardware store buying rolls of the stuff, which he'd use to make prototypes of his inventions.

"I took the wire I had, two water bottles, and a straw and stuck wires in and started producing hydrogen" is how he recalls it. It was during this experiment that he also produced his first disaster.

"I put a small hole in the bottle and lit it with a match," he says. "It didn't contain pure hydrogen, it wasn't a perfect hydrolysis model. It made an explosion, a pop noise, and I got splashed with a bunch of water and burst the bottle. I was wearing goggles."

He would later develop an obsession with wind turbines. "I had a new phase each month at that point," he says.

Even during his first few years of elementary school, though he could not have articulated it, he had an innate sense that he was different from his peers.

"I was like the idea man. There were a few kids who always wanted to dig things and I'd create tools for them. Kids wanted to climb this pole that only the big kids could climb and I'd figure out ways for them to do it," says Ethan.

By the time he was eight, his black-and-white marbled composition journal included ideas for companies, such as a survey company that bears resemblance to what are now commonplace online surveys. There are detailed diagrams for underwater power generators, as well as Ethan's notes on the flaws he perceived in various cars.

In an entry dated May 25, 2009, titled "Problem with Hyundai Genesis," he writes, "The problem with the Hyundai Genesis is that the grill is shaped in a way where it has harable [sic] cornering and the Hyundai symbol isn't in on its front grill."

He later details his thoughts on stock prices for his theoretical company, one of his most prescient entries. He even makes a run at explaining P/E ratio.

In middle school, Ethan had a business card for his first significant invention, which he dubbed "Ventricity." He built a tiny wind turbine that when placed over central heating or air-conditioning vents generates electricity. At the top of the card, it says, "Ethan Novek" and beneath, in slightly smaller type, "Inventor."

Below the word "Ventricity," the card reads, "Patent pending."

When Ethan joined Andy's class as a freshman, it was as if he'd finally found his natural habitat in a school setting. Though he'd always been a top student, the parameters of traditional public school education were tedious for Ethan. But walking into Andy's class, he discovered a sea of scientific instruments he'd never seen before, available for his use, as well as a teacher who would let him toil away to his heart's content.

"I was in the lab every day till six P.M., before school, weekends, anytime there was lab space," says Ethan.

Andy quickly saw that Ethan was boundless, in the truest sense of the word. Andy was more than happy to do what he could to help him.

Ethan has always been interested in all things energy, as well as anything wasted or underused. As a sophomore, he found a way to take waste heat—which, for example, would be produced by a factory—and create urea and electricity, both useful. At the 2015 Intel ISEF, his project won a first-place award and several specialty awards.

In the course of his research, he discovered the work of Yale professor Menachem Elimelech, an Israeli environmental and chemical engineer. Elimelech had pioneered a membrane-based power-generation technology that Ethan thought could be compatible with his work.

Ethan emailed the professor and got no response. He persisted, but still nothing. This irked Andy, who took it upon himself to send Elimelech an email urging him to give Ethan the courtesy of a reply. Eventually, Elimelech agreed to meet with Ethan.

Ethan and Andy drove to Yale to meet the professor in his lab. During the car ride, Ethan was busy keeping his expectations in check.

"We were kind of like in a sad mood," says Ethan. "Here I was, I had reached out to Meny and we're not expecting much from the meeting. People put realism in me, said, 'Don't get your hopes up. He's a professor, you be a good little kid.'"

Those who had tried to lower Ethan's expectations—for good reason—were wrong. It doesn't take but a few minutes for anyone to ascertain that Ethan is on another plane, both in terms of brainpower and maturity. Elimelech was no different.

What makes Ethan even more impressive is that he comes across as a completely good-natured, straightforward guy. You wouldn't be faulted for assuming a kid with this kind of intellect would be stratospherically dorky, but that is in no way Ethan. It's as if he's a thirty-five-year-old trapped in a teenage body. He has social skills, warmth, and self-assurance.

One thing he does do that betrays a certain lack of life experience is launch into conversations and explanations of his work with a full assumption that any audience can immediately dial into his discourse on aqueous ammonium bicarbonate and what happens when you add acetone, dimethoxymethane, or acetaldehyde at a concentration of 16.7 percent. He says these things as if he's asking if you, too, saw the most recent season of *Better Call Saul*. But it's the opposite of con-descending. He just assumes everyone knows this stuff.

So who better to understand and appreciate him than a professor who has devoted his life to such matters? Elimelech was quickly taken by the wunderkind, and possibly a tad humbled. Here was a fifteen-year-old suggesting he could incorporate Elimelech's membrane technology into something much larger.

While Andy was enjoying the "I told you so"–ness of the moment, he was hugely irritated when Elimelech turned to him and said, "Why are you here?"

To which Andy replied, "Where else would I be?" The idea that he wouldn't—or shouldn't—accompany his star student greatly of-fended him.

The exchange produced a sour aftertaste in Andy that hasn't faded since. For starters, Andy says he'd never send a kid to see an adult without some sort of oversight. Yale Schmale. Andy didn't know this guy and he wasn't about to let Ethan enter a meeting without his or some other adult presence. He had no concerns about whether Ethan could handle himself, but it made sense for Andy to be there, to vali-date that Ethan had done everything himself, that he hadn't been farmed out to a professional lab and wasn't riding someone else's coattails.

Andy sensed that the professor had no interest in or respect for a high school teacher. Of course, Elimelech likely had no prior knowl-edge of Andy's industry career, but regardless, his apparent disregard for someone teaching high school science bugged the hell out of Andy. While most of the time he can set aside his ego, this kind of disregard inflamed it. Andy never flaunts his corporate career to his colleagues because to do so would be to imply he was somehow superior. Which was why Elimelech's attitude tapped Andy's "fuck you" nerve.

Initial reluctance aside, Elimelech later agreed to allow Ethan into his lab to expand upon his work with the goal of authoring a paper to submit for publication in a peer-reviewed scientific journal. Two Ph.D. candidates initially worked with Ethan to make sure he was capable of operating in a university-level scientific lab.

So starting in the summer before his junior year, once he met the required age of sixteen to work in a lab, Ethan started commuting an hour each way to New Haven to work under the tutelage of Elimelech. He pressed on through the fall, making the after-school trek daily. But the commute and ensuing shortage of time to do his homework, eat, and sleep took a toll. His experiments were taking off, exceeding all expectations, including his own, and he wished to immerse himself fully without the quotidian pesterings of high school life. Ethan wasn't thinking about who was dating whom, he wasn't vying to be at the top of his class, and he had no interest in whether the basketball team won or lost.

No, his mind was fully absorbed by something else: carbon capture, currently one of the most exciting areas of environmental science. Carbon capture involves trapping carbon dioxide emissions and converting them into a usable form to make cement, carbonated beverages, etc. The difficulty is coming up with a process that itself doesn't require outsized amounts of emissions-producing energy, thereby reducing its benefits.

While Ethan was commuting daily to Yale, Bonnie could see that her son was stressed out and pulled in opposite directions. The family discussed their options and went to Ethan's guidance counselor and asked whether Ethan could be granted an exemption from attending school or doing daily homework. He'd make an appearance at GHS every few weeks for tests, and so long as he kept up his grades, the school would give him this dispensation. Lucky for the Noveks, Evan Dubin, Ethan's guidance counselor, was very much on Ethan's side. The deal was struck.

By winter, the Noveks decided to take one more step to further ease any unnecessary burdens on Ethan. They rented him an apartment in New Haven, within walking distance of the lab where he was regularly logging twelve-hour days. Bonnie would drive there once a week to clean the apartment, make sure he had food, etc.

She's the first to admit that this sounds completely bizarre— letting a sixteen-year-old live on his own and work at a university lab. "If someone asked me if they should let their kid do this, I'd say no," she concedes. But she and Keith felt Ethan could handle this unorthodox arrangement and they kept regular tabs on him.

But of course, as a mother, she questioned whether his premature departure from high school would be damaging to him in some way, that he'd look back later in life with regrets that his parents consented to this completely alternative path.

And it *is* completely bizarre—until you meet Ethan. When you probe for signs or even the faintest whiff that any of this is harming him, you come up empty. And Ethan hasn't completely left GHS in the rearview mirror. He plans to apply to Regeneron, the seniors-only competition with the top prize of $250,000, and he'll also enter CSEF.

If anything, it would be easier to imagine that forcing him to do the standard march through high school is the thing that would stunt him. When people say, "High school isn't for everyone," they mostly refer to recalcitrant kids who can't function in school due to any number of problems.

But high school really isn't for Ethan either.

19

Sophia

It's November 3 and Sophia's in the lab, prepping her trays to grow her biofilm. It's a precarious business, as Sophia discovered the previous year.

Once she'd formulated her project in tenth grade, she set about growing her Lyme biofilm. Believe it or not, there are recipes online to do such a thing. The tricky part is trying to create the precise conditions that allow the biofilm to grow and prosper while avoiding any kind of contamination. Several of the samples Sophia had cultivated last year somehow got spoiled. When asked how you can detect such a thing, Sophia's eyes grow wide and she says, "Oh, you know." Scary-looking sci-fi-grade matter appeared in her trays. But it's impossible, short of spilling an obvious contaminant, to know precisely what caused the defect. It's as if there's an invisible grim reaper constantly lurking and threatening to sabotage the whole thing.

Last year, Sophia prepped her trays in the open air of the lab. She would assiduously swab down the tables with alcohol before beginning her work, but still the contamination crept in. To give you an idea of the level of sterility used in a professional setting, biofilm is grown in what are called clean rooms, sterile spaces where scientists wear full body cover, shoe covers, masks, etc.

To try to stave off any damage this time around, Sophia has decided to grow her biofilm in one of the lab's glove boxes, a giant glass cage of sorts with long rubber gloves in the front that extend into the box so you can work in sterile environments, or with hazardous materials, while reducing risk of unwanted exposures.

The very first step for growing biofilm stands in direct opposition to Sophia's animal-loving nature: she has to defrost some rabbit blood. Fortunately, Sophia, a pet collector who's managed to cajole her parents into two dogs and a one-eyed cat, has never had a pet rabbit. Andy asks the kids to write up their materials lists and then he spends hours on the phone, usually with Sigma-Aldrich, a chemical company, because the stuff has to be ordered by an adult and shipped to the school. He likes it when there's some overlap and kids might be able to share materials—nanoparticles, solvents, or some such—to keep costs down.

Sophia, however, has the distinction of being the sole kid in need of rabbit blood, which is what the recipe calls for as a stand-in for human blood. Once defrosted, it has the color and consistency of cherry Kool-Aid, much thinner, brighter, and more translucent than human blood.

Sophia unstacks her trays and lays them on the table. These well plates, as they are called, are about six inches long and four inches wide. There are three horizontal rows with four recessed circles about an inch deep, slightly smaller than a quarter, which will house the chemical concoction intended to yield the biofilm. Sophia pipettes five milliliters of bunny blood into each circular well, aiming to make the layers as level as possible.

Next she loads her pipette with liquid agarose, a polymer that will give the mix a gelatinous consistency. After pipetting the agarose into the rabbit blood, she has to wait ten minutes for the mix to firm up. She notes the time and heads to her laptop to do some work.

Andy's kids often do continuations of projects, spending two years elaborating on and extending an original idea. The previous year, when Sophia made her first attempt to grow her biofilm, she administered the PDE4 inhibitors and showed that they were effective in destroying the biofilm. But it was time-consuming and labor-intensive, especially at the end when she had to slice her samples in a professional lab where they then had to be viewed under the SEM, the ultra-high-power microscope. The time involved in getting to Mount Sinai to use the SEM was such that she missed the state fair her sophomore year. Still, she took her preliminary findings to the Phenomenon: Science Innovation Fair, a local competition in Greenwich, and won the People's Choice Award. But this year, her goal is to complete a three-step process of growing her biofilm, adding the PDE4 inhibitors, and then administering several combinations of antibiotics to the mix to see whether they have an effect.

After ten minutes, the rabbit blood/agarose layers have formed a gelatinous, pinkish substance the color of cherry Jell-O.

There's a nascent buzz in the room, as kids are now starting to delve into the hands-on aspects of their projects. Manuel "Noni" Lopez, a junior, has started his project in earnest—a diagnostic test for Chagas, a bug-borne illness prevalent in his native Venezuela. Olivia is working on retooling her Ebola project for Lyme disease, so that her paper-based test can detect the disease through a saliva sample, as opposed to blood serum. Like so many kids in Connecticut, Olivia has known many people felled by Lyme, which made her interested in improving diagnosis.

The kids work on their own but pivot to Andy for consultation on next steps. As a first-timer, Noni is eager and jumpy, flitting around the room with his chemicals.

"Gloves?" says Andy. "Goggles?"

"Oh, right," says Noni, running to suit up.

As for Sophia, she's now got her jar of bacteria on the table. Science fairs have stringent rules and regulations around what kinds of materials kids can handle, as well as the bounds of their experimentation. For obvious reasons, kids aren't allowed to do any invasive, chemical, or medical testing on humans. Andy has limited animal work done in his lab to fruit flies and mice. For the mice, his student performed an observational behavioral study. There was no cutting or injecting.

But all the regulations mean Sophia cannot work with the active bacteria that causes Lyme disease, *Borrelia burgdorferi*. Instead, she uses a safe, nonpathogenic form, *Borrelia turcica,* which is biologically similar to Lyme bacteria. Andy located a company in Germany that shipped the bacteria to Greenwich High in a clear plastic jar.

What Sophia is doing is essentially building a multilayered sandwich of chemicals in each well. The next layer is more rabbit blood/agarose, but with different proportions than the initial one. She's now placing a smaller amount of agarose on the top. Finally, she pipettes the key ingredient, the bacteria, a yellowish liquid, into the wells.

Sophia's filling three trays because she's going to be looking at several combinations of interventions: three different types of PDE4 inhibitors alone, PDE4 with tetracycline, PDE4 with doxycycline, PDE4 with the combination of the two antibiotics, and antibiotics alone. And of course, she needs controls, untouched wells to which she does nothing in order to see how they behave by comparison.

The recipe calls for a 5 percent carbon dioxide environment in order for the biofilm to grow. This can be created by lighting a candle in an enclosed container. Once the candle naturally extinguishes, a 5 percent CO_2 atmosphere has been reached.

As she was leaving her house for school, Sophia grabbed one of her mother's periwinkle blue pillar candles to bring to the lab.

She opens the top of the glove box, places her three trays inside, lights the chunky blue candle, and shuts and fastens the door. The biofilm can take weeks to grow. The fancy candle cuts a funny figure in the glove box—Pottery Barn meets mad science. The sacrificial candle glows brightly, creating the ideal chemical ambiance in which to grow biofilm.

Looking at it, Andy smirks and says, "Ah, it's like Christmas."

— — — —

On November 17, a week before Thanksgiving, Andy writes a message on the whiteboard at the front of the lab:

Dec. 1: CSEF app deadline
Dec. 21: CT STEM app deadline

Start now plz!

Since his class has no homework assignments, quizzes, or tests, Andy does very little in the way of housekeeping and nagging. Science research is an honors class at GHS. Andy gives quarterly grades based on effort and formal project updates, usually in the form of a PowerPoint presentation. He bypasses daily badgering in favor of the occasional stark notice on the whiteboard.

When the kids walk in and see the otherwise empty board bearing a message, they all instantly go into slow motion, staring at the message while removing their backpacks. Andy uses his whiteboard so judiciously, the kids take the missives seriously.

But the next day, Andy decides that the CT STEM Fair merits a classwide discussion. He sits at one of the lab's back tables, laptop open.

"All right, everyone. Woohoo. Can I get your attention?" he says. It takes a minute for the kids to swivel around and focus on him, especially since he sort of blends in with the crowd while sitting down.

"So, as I told you guys a few weeks ago, I got that call about CT STEM now having ISEF affiliation. In the past, we've never gone to CT STEM because it's a smaller fair and since there were no big prizes, we focused on CSEF. But now that there are these three ISEF spots, I want to encourage as many of you as possible to apply.

"What does this mean? This means that you have to get your projects done a month earlier because this fair is in the beginning of February.

"Can you apply to both? Yes. And if you only wind up applying to CSEF, and by the way, for some of you, given the timing, that might be your only option, that's fine. But I'd love to see you guys take a swing at CT STEM. I've been there, I've judged that fair, I know the caliber of what's there."

He pauses.

"I'll be honest. I want you guys to go out there and win this thing."

Listening to Mr. B., Sophia knows she wants in.

Over Thanksgiving break, Sophia and her dad, Gregg, fly out to California to pay a visit to UC Berkeley, where both Gregg and Stephanie went as undergrads and where Gregg went to business school. Saturday, November 26, is the final football game of the season. Berkeley is playing rival UCLA and the Chows have been invited to sit in the Chancellor's Box. The Golden Bears are having a bruising season, having lost the previous four games.

Sophia's a bit leery about all the college stuff, the ultimate pressure point of high school in this day and age. She lovingly calls Gregg "Tiger Daddy," as he's the academic czar of the household. He lines up tutors and keeps track of the kids' grades. Stephanie says her role is to be the cheerleader, the supporter, since Gregg oversees the logistics and monitors their performance and extracurriculars. He's found a streaming service that allows him to watch all of Brandon's baseball games from his desktop at work.

What Gregg is doing is giving his kids what he never had. He was fifteen years old when his father, Howard Chow, a NASA engineer, died of lung cancer, ripping a huge hole in the family. His older brother, Blake, was just finishing high school. His mother, Marilyn, was an elementary school teacher in a rough part of San Jose.

Without question, Gregg says, the loss of his father in his early teens became a flashpoint that shaped his work ethic and sense of purpose. He harnessed his grief over his dad's death into running and became a track star, as his father had been. (Thirty-five years since he graduated from high school, Gregg's still ranked number 72 in the

California Central Coast All-Time 100 for the one-hundred-meter event.) He jokes, "Fastest Asian, I bet!"

He also started dating the most beautiful girl in the school, Stephanie Clarke, whom he'd marry in his twenties.

After graduating from Berkeley, Gregg worked as a programmer and systems engineer at Chemical Bank. He excelled in this arena, producing a novel software program at age twenty-three for foreign exchange trading. He then went into market analysis, eventually making it onto the trading desk. Today he's a managing director at a major investment bank overseeing a sales force that covers global investors.

Gregg says he's spent fatherhood mulling the following question: "How am I going to instill hunger in Sophia and Brandon without dying? How do I do this—in Greenwich—without dying?"

Because even without the early loss of his father, his upbringing—and Stephanie's—stand in sharp contrast to that of their kids, who wake up each day in their plush waterfront home with startlingly beautiful views of the Long Island Sound. They've had a life of lovely vacations, of meeting famous people and growing up in a hell of a nice cocoon.

With their California sensibilities, Gregg and Stephanie seem to take a somewhat bemused view of Greenwich and its outsized affluence. Despite having nothing in the way of family legacy to lean on, they've integrated socially. They belong to Innis Arden, one of the oldest country clubs in town. (Back when he and Stephanie applied, Gregg was initially pessimistic about their chances of getting accepted, country clubs not exactly being historical bastions of diversity. He has a certain pride that his class of new members at the club included him, a fifth-generation Chinese American, and a close friend who's Jewish.)

But the Chows tell Sophia and Brandon all the time that they're growing up in a bubble, that Greenwich is in no way a reflection of reality, that once they leave home and see how most of America lives, it's going to be quite eye-opening for them.

Stephanie says, "I tell them, when you go to college, the friends you make are probably going to be shocked when they see how you've grown up and you might be shocked how they've grown up."

Gregg would love for Sophia to attend Berkeley, if Berkeley's the kind of school that appeals to her. And on game day, how can it not?

She and Gregg are down on the field during pregame practice, within inches of the players, tanks posing as humans. There's so much revelry and excitement and a stadium full of school spirit, awash in gold and blue. And as luck would have it, the Golden Bears beat UCLA 36–10, a fist-pump finish to a lackluster season. Gregg's Bears spirit gets a nice jolt, and when Sophia comes back home and describes the thrill of the game, it's clear she's smitten.

While father and daughter Chow are in Northern California, Gregg takes Sophia to visit UC San Francisco, UC Santa Barbara, and Santa Clara University—none of which really speak to her. UCSF's too urban for her and Santa Clara's "really clean," but small.

They conclude their trip by heading to San Jose to pick up Bobby Boo, which is what Sophia and Brandon call Marilyn Chow, their eighty-four-year-old grandmother. Sophia bears a strong facial resemblance to Marilyn, a petite Chinese-American woman who wears skinny jeans and loves Facebook and texting. Gregg and Stephanie both recall that upon seeing Sophia on ultrasound when she was in utero, Stephanie exclaimed, "Oh, my god! It's Marilyn!"

Sophia and Gregg are bringing Bobby Boo back to stay with them for the holidays in Greenwich, where the homes are trimmed to tasteful perfection. There are very few, if any, inflatable lawn Santas in this town.

On December 1, at the start of the holiday season and fresh off her California jaunt, Sophia walks into the lab and heads to the glove box. She removes the trays. What she sees is something she's never seen before. The samples have sunk low into each well. They're hard and waxlike.

There's nothing in the way of biofilm.

Something has gone very wrong.

20

Romano

Following the great breakup of 2016, Romano gets some modern-day youth blowback. Immediately after the incident, he's bombarded with texts from Camp Caitlin telling him he's a loser, an asshole who will regret breaking up with such an amazing girl. Romano finds it all very tiresome and is a bit incredulous that people would invest time in assailing his character over a breakup that has nothing to do with them. Then again, this is high school. And one of the more pithy conversations he and Caitlin engaged in was about the tragedy that is peaking in high school—that to do so was to tilt life's ramp downward. As entangled as they could get in the spidery web of teenage dynamics, they both knew it was wise to look beyond these years, as abstract as that may be.

Yet, in the aftermath of their breakup, a tidal wave of pettiness has come crashing down. He doesn't respond to any of the taunts, which seems to declaw them, or at least not spur more.

Despite it all, Romano can now turn his attention in full to his research project. With neither girlfriend nor looming breakup hanging over his head, he feels an immediate sense of liberation. He even says, tongue fully in cheek, that now he can go back to being "a stallion."

On Friday, November 11, the kids' research plans are due—detailed blueprints for their projects. Romano spends the weekend before reading as much as he can and consulting with Bruce Lee, the expert on L-DOPA.

Andy ordered the L-DOPA for Romano, but Romano is still trying to nail down the process he'll use to create his liquid bandage. He has a perfectly formed vision for the end product, but his thoughts on how to get there are thready and tenuous.

Here's what he knows: L-DOPA is a natural sticky amino acid. In theory, it should bring adhesive properties to things that otherwise would be without them. But L-DOPA alone can't and won't act as a liquid bandage. So what should serve as the base? And what can Romano add to give it some additional healing properties? These are the great unknowns.

So on Monday, November 7, Romano walks into the lab with purpose. He takes off his purple backpack and removes his opus, the research plan.

Andy takes a seat at Romano's table. He sits with Collin Marino and Joe Konno, both sweet, introverted kids with braces and big ideas. Collin thinks very abstract thoughts about cellular behavior and is working on how to program RNA to essentially assassinate cancer cells. Joe, a highly accomplished violinist, is working on a project that could have implications for hepatitis C. They exist in an entirely different social sphere than Romano, but the three somehow always manage to find something to banter about, be it Star Wars, videogames, or what really happens at Greenwich's weekend parties—the third being something that, at this table, Romano alone can provide the answer to.

Andy's now reading over Romano's paper and a long discussion ensues. Romano's eager to show Andy that he's been devoting a good deal of mental real estate to his project.

"The reason why a mussel can adhere to a rock underwater is because of L-DOPA," says Romano. "The reason why this liquid bandage will be able to adhere to a bloody surface even after taking a shower is because of L-DOPA."

This is Andy's first deep dive with Romano on his project.

"I want to try and keep this as natural as possible," says Romano. "A perfect starting point is gelatin." He's been tossing ideas around with Bruce Lee, weighing the possible benefits of various options that will serve as the base of the bandage. Bruce Lee has worked with expensive substances called PEGylated amines that yield soft, versatile gel-like properties. He steers Romano away from the PEGylated amines, both because Bruce Lee himself has already done this and because of the cost.

Andy's not immediately deterred. He starts searching and it looks like they cost, at a minimum, about six hundred dollars.

"That's more than my Xbox," Romano cracks. "You think I'm selling my Xbox for this class?"

Andy's wondering whether a hydrogel would work. A hydrogel is a squishy mass of soft, polymeric beads that can be manipulated for a variety of medical uses, such as encasing drugs that get swallowed and even . . . bandages. Some of his kids have had great success with hydrogels.

Romano says, "What I'm thinking my goal could be is if you trade out the normal plastic bandage for a Band-Aid that you can, like, squeeze out of a tube and then it solidifies in, like, five minutes . . . this is my ideal project. Solidifies in about five minutes, denatures after about two days, the L-DOPA is what will keep it sticky, the polymer is what will be the base, and then you can, like, add stuff to it. If I want to keep it all natural, I read about the antibacterial properties of lavender oil and frankincense and stuff like that. I mean, essential oils are kind of shady, the science—"

"You think?" says Andy, eyebrows up in sarcasm.

"Once I have something that can stick to the skin, is nontoxic, and has a base, you can add whatever you want to, pretty much," replies Romano.

Not so fast, partner.

"Well, it has to be absorbent and release it," Andy says. "I mean, they talk about hydrogels in here, but it looks like when they were doing it, they were taping it on. It doesn't have adhesive properties on its own. Maybe your take then is to make an adhesive hydrogel that can deliver meds."

"Is there any way to keep it natural? Is there any way to keep it organic because L-DOPA is organic?" Romano is a bit obsessed with the "all natural" aspect to his work. Laudable, but not quite practical.

"Hydrogels aren't natural," says Andy. He gets pulled away by some questions from other kids needing his attention, and the conversation at the table turns to *Game of Thrones*.

"Who raises such a kid?" Romano says. He's talking about Joffrey Baratheon, the sadistic teen prince turned king. "Such a Greenwich kid. He'd fit in here. So annoying." And speaking of drama, Romano has landed a part in the school winter play. He's taking a comedy and improv elective and his teacher urged him to audition.

This is yet another whole new realm for the former football player turned science researcher and now thespian. The play is *Noises Off*, the 1980s Michael Frayn farce that has been revived again and again on Broadway. Romano has been cast in the role of Freddie, a lovable dunce. The role calls for sight gags aplenty, with lots of underwear reveals. Given how uninhibited Romano is, the casting seems on point.

There's just one thing: Caitlin has been cast as one of the female leads in the nine-person show. Romano can't quite catch a break. It was bad enough that the Monday after he broke things off, who does he run into on one of the staircases in the student center after school? Her parents. In a 450,000-square-foot school of nearly 2,700 kids, what are the chances of running into your newly ex-girlfriend's parents? He says her mom gave him the stink eye and he just kept walking, his heart thumping. He was in complete disbelief that he ran into the two people who most likely wanted his head on a stick, *Game of Thrones* style. He didn't downplay the fright of the encounter, saying simply, "I was gonna poop my pants, dude."

Andy's now back to Romano and the L-DOPA quandary.

"What do you think?" asks Romano. "Adding other stuff? Like, lavender oil? I just threw it out there. Or adding an antibiotic?"

"That's an option. It's a little more credible. Leave the oils out of it," says Andy. Romano says he'd like to make his own polymer base for the bandage, while Andy's noodling another option.

But Romano is fixated on L-DOPA. "If we end up creating our own polymer to polymerize L-DOPA with, I think that could actually be pretty good because then I have to sort of refine the ratio of ingredients to the polymer to achieve ideal properties," he says. Even if his thinking isn't quite there yet, he's dropping terms like "polymerize" and "ideal properties" like a baller.

But Andy's on a different train. He's deep in Google, plugging in various things to see if there's a prefab recipe for a hydrogel bandage.

"Okay, here's a paper, recent. It talks about an amorphous hydrogel for wound treatment because it has silver in it. Silver's antibacterial. So what you could do is make this hydrogel with silver and then make it sticky."

Andy is triumphant. "That's it, man," he says. "Here's the recipe on how to make it. We could probably do an antibacterial test. I think this is where we should probably jump off. If you didn't want to do silver, you could put drugs in there."

Romano's face is the picture of skepticism. Something about this isn't sitting well with him.

"Because right now," says Andy, "if I can be frank, you're in love with this chemical, but you don't have a real use for it yet."

"Which chemical?"

"This binder, this L-DOPA. I mean, I'm sick of talking about it. I want it to start moving."

"Yeah, I can agree with that," says Romano.

"So, why reinvent the wheel? The novelty for you is to create the adhesive aspect of this," says Andy. "Why not take something that's been made before, someone's going to tell you how to make it, you got the proposed benefit for it already.

"You're altering the composition and making it sticky. But you don't have to sit and wonder, how am I going to make this polymer that's going to work best? They tell you how to make it.

"But you have to be clear, as you highlight this work, should it go till the end, that you didn't come up with this design for silver nanoparticle hydrogel. But you saw the inequities and limitations of it because it falls off as soon as you put it on. And why carry a bandage? Wouldn't it be nicer to carry a tube where you pffft, squirt it out? It sticks and you keep hiking. Make sense?"

"Yeah, it makes sense," concedes Romano. But he has a look of utter deflation on his face, as if someone has just stepped on his bag lunch.

"It's what I do, man," Andy says. "It's a line from *White Chicks*. You ever see *White Chicks*? You've never seen *White Chicks*? You?" Andy's polling the table looking for anyone who's seen the movie so they might appreciate his use of the line.

"That movie is funny. Not to be watched with your daughter," says Andy, adding, "the TV version is actually quite clean." (Andy's bible of funny consists of *White Chicks* and *Superbad*.)

Finally, Romano comes out with it.

"I'm honestly afraid that this isn't *mine* enough. I'm just taking this recipe and shaking a little molecule on it."

"No, no. I'll play devil's advocate," says Andy. "As we sat here and we talked, we said we need a hydrogel that's going to be a wound hydrogel, so we start spinning and then it turns out that this amorphous hydrogel is the magic hydrogel, not new to this paper. And we need to pack something in there that's got function.

"Just keep in mind, what is the benefit that *you're* showing? What are you about? You're about the adhesive part of it. So you don't really have to sit on this. This is where you jump off. So would it be better if you designed it from A to Z? Sure. Do you have the capacity to do that? No. We don't. And there's no shame in doing it this way, there really isn't.

"Everything in science is jumping off a previous paper. I mean, silver, we've used silver a lot. We have antibiotics, we have them in the fridge from previous kids, we have amoxicillin. Want to put that in there to make it different? Sure.

"But at least you have a framework from which to build that hydrogel. And if you sat there and said to some judge, I designed this hydrogel, they'd laugh in your face. Hydrogel's been around. There's no novelty there. It's the application and it's the improvement in its function," Andy explains.

Herein lies a lesson about science that Romano could do with hearing. He's arrived at the class thinking that for his work to be authentic he needs to personally construct every single aspect from scratch. It's as if he's tempted to don some snorkeling gear, collect mussels, and extract the L-DOPA himself. He doesn't realize that science is a discipline that constantly and necessarily builds on preexisting work—which doesn't in any way undermine ingenuity and new discoveries.

"I understand," Romano says, perking up a little. "Thanks for saying that. One last question: I understand this isn't all natural. Is there any natural hydrogel or any natural polymer? That's the only reason I was leaning towards gelatin so much is because that's like—"

"I'm just afraid gelatin's going to fall apart," says Andy. "When I think gelatin, I think Jell-O, and when you put it in water, it becomes like something you don't even want to wash in your sink. It's gross, it clogs sinks, doesn't it?"

"That's what L-DOPA is good for, it keeps it, it binds it," says Romano.

"Well, is it binding it, making it adhesive? I mean, that's a difference, right? If you took a piece of Jell-O that's not falling apart and you stuck it on your arm, guarantee it's gone, right?"

"What's the endgame is really what you have to decide," Andy continues. "Is it to make a really low cost natural gelatin material that you have to tape on? That's okay. Or is it that I want to make this really cool hydrogel that is pretty common, people are using them, but I'm going to load it with some stuff *and* I'm going to make it stick?"

Romano says nothing so Andy keeps going. "Right now, it's just a friggin' booger that falls off, that's really what it is. It's a booger that you've got to tape down. That's okay, but, man, I think it would be a lot cooler if you could make a product that just squeezes out of a tube."

"Can I do both at the same time?" asks Romano.

"No, no. Now you're dreaming. Not with your play. And going down Main Street in convertibles. No. Could you? Sure. Should you focus on one? Focus on one and see where it goes. I don't think this is bad, I really don't."

There's a pause in the chat while Romano mulls this over. Collin sees his window to grab Andy.

"Mr. B., do you think you can read my methodology real quick?"

"Real quick? I've been teaching for six hours, dude. Yeah, yeah, when I'm done with him. If you want to print it out, I'll take a look. Joe, anything from you?" Andy's checking for other takers.

He turns back to Romano. "So what do you think? I sense a little . . ." Andy's scanning Romano's face.

"I don't know. Just hesitance, that's all it is," says Romano, his tone still like that of a kid on Christmas morning coming to terms with a gift that's close, but not exactly what he wanted. Like getting a PlayStation when you asked for an Xbox.

Andy sees the dejection.

"The problem will be if you just start buying these PEGylated amines, they're expensive. If you want to try something . . . you're thinking of gelatin? We'd have to make Jell-O, put some L-DOPA in it, and see if . . . we could try that. We could do whatever you want."

"Maybe all I need is a taste of the bad so I know what's good," says Romano.

"Go eat some pizza in the cafeteria," retorts Andy. "There's a taste of bad."

"What if we tried, first off, with the gelatin and if it's a complete bust, then we—"

"You don't have a lot of time," cautions Andy.

"I don't have a lot of time?"

"No. You've got less than a month. Today's the seventh. You've got to get this thing nailed down to apply to the fair."

"Which fair?" asks Romano.

"State fair. If you want to give it a try, then you should be busy this week playing with Jell-O. So if you want to get some Jell-O tonight, we have the L-DOPA, right? You could mix some Jell-O, put some L-DOPA in it, and see what happens. We can try that, sure."

"All right. Thank you, Mr. B." Romano's shed his wilted daisy face and looks genuinely pleased that Andy's willing to let him venture down the path of his choice.

"What are you thanking me for? It's just an idea bounce, that's all. That's the reality of it. The wheel has to start rolling down the hill."

"I personally think *this* is an option," Andy says, pointing to the paper on his screen with the hydrogel recipe. "It's published in 2015 so it's brand spanking new and so it has novelty."

But there's no turning back. Romano's all about the gelatin and he's not letting go. Andy opened the door and Romano rushed in.

"Go buy me some Jell-O tonight. No color," says Andy.

"No color? Can I get, like, strawberry or something . . . ?"

21

William

On the north side of I-95, behind a McDonald's and a strip mall, is an unremarkable neighborhood that could be anywhere in America. Technically, this neighborhood is part of Riverside, which is considered to be one of the swankier parts of the Greenwich area. But this section of Riverside isn't posh. It has a jumble of homes built in the 1950s and 1960s and few, if any, have been updated. By the look of their exteriors, it's easy to picture shag carpeting, wood paneling, and Corningware dishes within.

The Yins live in a modest 1,614-square-foot, four-bedroom, one-and-a-half-bath split-level with yellow siding. Their small living room is dominated by the piano played by William and Verna. As is customary in Asian households, shoes come off at the door. The house is immaculate, not a speck of clutter to be seen (until you get to William's room). The family cat, Mars, adopted at the behest of the kids, slinks around the place.

Harvard's early admissions decisions are landing at five P.M. today, December 13. The verdict is no longer indicated by the envelope size: big and thick for yes, thin and spare for no. Applicants now log on to a portal, which delivers the answer in cold, swift fashion, but with the same digital sangfroid to which kids these days are accustomed.

The Yins are anxious and a bit giddy with anticipation over the decision. William had what he thought was a good interview with the Harvard alum (unlike his Stanford interview, where the alum had a written list of questions on a legal pad and refused to allow for free-flowing conversation). Dr. Winters wrote a thoughtful recommendation, where he tried to highlight unexpected things about William. He was conscious that on paper William was at risk of coming across as a cookie-cutter Asian high achiever, the science-minded, piano-playing, perfect-standardized-test-scoring kid. So instead he wrote about the reach and randomness of William's pursuits, like playing in the pep band at football games and going to his junior prom—the latter being an odyssey that involved Andy.

Being a seasoned guidance counselor, Ms. Patti knows her best play in regard to Harvard expectations for most kids is blunt reality: there's no gain in offering even a wisp of false hope. But since William is "truly remarkable"—the top student she's had in her career spanning nineteen years at Greenwich High—she told him there's no reason to think he won't get in. Plus, Harvard's regional admissions officer is said to be a big advocate for William.

So when Mark, William, and Verna (Wendy's not yet home from work) gather at the kitchen table a few minutes before five P.M., it feels like a garland of good fortune has been draped around the house. Their Christmas tree is up, and the multicolored lights are shining.

The computer William built in middle school resides beside the family's dining table. The tank of a hard drive sits on the floor and the monitor is on a cabinet that straddles the living room and kitchen.

William's wearing his gray GHS tennis hoodie. He pushes up his sleeves and moves the keyboard and mouse pad onto the table, smirking as he points out that the mouse pad is from MIT. He's created a countdown clock and the screen reads, "It's almost decision day."

At 5:00 P.M. sharp, he checks his email. Nothing. He pulls up the

Harvard portal. He enters his log-in and password. The screen is mostly blank. The next time he checks, he gets a screen with generic contact info. He refreshes the page several times, tries logging in again and again.

Mark has his phone trained on William, recording the moment. But after a dozen tries, he puts it down.

"I mean, it might not even be by five," says William. "Maybe six."

"Midnight?" Mark asks.

"It wouldn't be midnight because then that would be tomorrow," says William. "But maybe eleven or something."

With each passing minute of delay, there's a growing sense of tension in the room that's extinguished any excitement.

William sets up the computer to auto-refresh, sparing himself the repetitive action of logging in. Mark and William question whether today is in fact the day, but all the other kids who applied early to Harvard believe it's today as well. William takes off the auto refresh and resumes the manual effort. The tension has now passed. Everyone seems to have exhaled, and a bemused bafflement takes over.

Finally, at about twelve minutes after five, William refreshes the admissions portal and a screen of lengthy text appears. The answer is in.

"Ah," he says, less than a second later. "Deferred."

"Oh," says Mark.

"Oh, well," says William.

"Huh," says Mark.

Verna says nothing, but she doesn't have to. The family's emotions are pooled into her eyes, small wells of shock, bewilderment, and sadness, eyes that are at once a scream and a cry.

Wendy walks in with boxes of pizza, all smiles.

"Deferred," says William.

"Deferred?" asks Wendy, more puzzled than disappointed. "Oh." She sets the boxes down and disappears for a full thirty minutes.

William, Verna, and Mark eat their Domino's pizza in small, mouselike bites, enveloped by a silence that is achingly painful. Their cultural sinew is such that you don't openly emote, you accept. (William says that his family doesn't say "I love you" to one another. "It's fully implied, we just don't say it," he notes.) This mindset, absent of any privilege or sense of entitlement, mutes a moment that in other Greenwich families might play out with indignation and immediate calls for action. Here, the added element of immigrant mentality—the lie low/be grateful you're here/don't draw any kind of notice modus operandi—is playing out in excruciating fashion. The tacit and automatic resignation, even while the moment is still hot to the touch, feels unjust, because it's both so seamless and unchallenged.

If ever there were a moment to invite some outrage, to blow the lid off the tightly sealed can of politesse, it would be now. But the Yins continue to eat their dinner silently, eyes glazed over in disbelief.

In an added twist of cruelty, William must go play his French horn tonight in the holiday concert.

The family slowly thaws from the initial shock and begins to engage in a dissection of what went wrong, seeking to understand how this destructive meteor landed in their backyard.

Mark focuses on the next tangible step, which is that William must now file a slew of regular decision applications. The idea that he might be accepted to Harvard regular decision is batted away with immediacy by William. He insists that a deferral is just a no by another name.

"The admission rate off an early deferral is like three percent. This is basically a rejection," he says. He picks up his phone and starts seeing what other kids are hearing. A few acceptances, but many deferrals. He seems to find some comfort in hearing he's not alone.

"So now you have to focus on the regular decision applications," Mark says. "You can't wait. Don't wait until the last minute."

Wendy reappears and seems stoic and disappointed. There are no hugs or tears, or even hands on shoulders. William doesn't seem to want that and his parents don't seem inclined to give any of it.

Wendy wonders out loud whether William was foiled by the Harvard alum who interviewed him. She points out that, coincidentally, the alum is a bank customer who always looks unhappy and seems like a "miserable person." Was she the source of William's undoing? Perhaps, yes, says Wendy.

Verna speaks up to dismiss this suggestion of conspiracy. Very slowly, haltingly, she says, "I . . . don't . . . think . . . she . . . sabotaged . . . William."

At one point, after another period of silence in which everyone seems to have folded inward to have their own think about the situation, William gestures at the Domino's boxes and declares, "Danny would hate this pizza."

Anyone could argue that it's foolish to expect your child to get into Harvard, that statistically the odds are impossibly against you. And Mark has always been conservative, maintaining more of a "you never know" approach.

But still.

In his class of 650 kids, if William is not academically number one, he has to be damn close. (GHS doesn't rank students until a few weeks before graduation, and even then, only the names of number one and number two are released for the purposes of naming valedictorian and salutatorian.) The fact that he excels at nearly everything and is a person of so many dimensions, that he has a perfect ACT score of 36, that he's bilingual and taught himself to code, that he developed a sticker to detect atherosclerosis . . . the list goes on.

So it is with no surprise that the next day, the GHS administration is stunned by William's deferral. It feels like a personal wound to those who know him, have admired and mentored him.

"Fuck them," says Andy, letting his Bronx flag fly. "If they're too stupid not to take you, you don't want to go there anyway."

William has a postmortem with Dr. Winters, who says, "I thought for sure William would be in the top five of their early admissions pool." Meaning, of the ten Greenwich High kids who applied early to Harvard, he assumed William would be at the top of the pile.

Ms. Patti is equally shocked. She immediately picks up the phone to dial the regional admissions officer (who can advocate for kids and make recommendations but does not have the final word) to try to get some intel on where things fell apart for William.

For his part, William is equally intent on figuring out what happened, and he tosses around everything from the Greenwich wealth stigma (which doesn't apply to his family) to the favoring of legacies. Ms. Patti tells him that for the past few years, early admission spots given to Greenwich students have gone exclusively to legacies and athletes. Three years ago, eight GHS kids were accepted early to Harvard; six were legacies and two were athletes.

William has to pick himself up and devote his energy to working on the regular decision applications. The Yins call off their holiday vacation plans, not even venturing to New Jersey to visit with family friends.

He walks around in shock for a few days after the news, which ripples through science research like an earthquake. Many of the kids take it as a scary omen about their own college admissions. If Yin doesn't get into Harvard, all is not right with the universe.

22

Olivia

December 15 is frigid, with temperatures in the low twenties, but inside the Guggenheim Museum on the Upper East Side of Manhattan, it's borderline stifling. The museum is the site of STIR, a peculiar conference that bills itself as a "dream-like gathering" and an "un-conference." Its description says "No breakout sessions" and "No business cards." The most specific STIR gets in its mission is to say that it's "STIRRING up what's possible in health care." But in case you need something more concrete to understand what the hell this gathering is about, the site says, "Imagine TED meets making health care better for the change agents seeking inspiration and ideas to heal health care—in a day." That's an awkward, squishy sentence attempting to describe an awkward, squishy meeting.

The un-conference gives out notebooks and small colored pencils for attendees to sketch, take notes, or reflect. There's even a clutch canvas pencil case to hold your inspiration materials. A live band plays soft, twangy ditties between each speaker. There are moving clouds being projected on the ceiling and the colors and logos are rendered in soft pastels. It's all very precious and ethereal.

STIR charges $895 for the day (which includes a dinner of post-modern, geometric food and guided tours of the Guggenheim)—and as in many tone-deaf, elitist conferences, the speakers talk about making the world a better place for others, specifically for the vast legions of people unrepresented at today's gathering. You can scan the room and quickly count the people of color, who are here in glaringly small numbers. Attendees are largely from corporate healthcare companies, and looking at the staid sea of suits, it's hard to imagine this crowd doodling with their colored pencils or coming up with fixes to the multipayer system based on gazing up at the cloud projections.

The roster of speakers is puzzlingly random, with no thread or tangible organizing principle connecting them. Poets and neurosurgeons and consultants. Some are regulars on the public speaking circuit. Some have ties to science and healthcare, but then there's a guy whose speech is a rallying cry for us all to think more like children. Not an unadmirable pitch, but probably not going to transform the very grown-up morass that is healthcare.

The youngest speaker is Olivia. She's been invited to talk about her Ebola work, of course. In introducing her, a lady in a black cocktail dress says she was recently "named in *Wired* magazine as one of President Obama's favorite young scientists. Please help me welcome the brilliant Olivia Hallisey."

Olivia walks out onstage looking strikingly older than she does on any given school day. She's in high heels and a long-sleeved forest-green dress. The STIR folks have done her hair and makeup.

"The power for science to make a positive change in people's lives has been apparent to me ever since I was a young girl and would visit my grandfather, a doctor and medical researcher," she says in her opening remarks. While she speaks, projected behind her is a sweet black-and-white photo of her grandfather holding and kissing Olivia as a toddler.

Olivia then takes the audience briskly and concisely through the Ebola epidemic, as well as her project—using a slick slide presentation. She's a masterful public speaker. She has no notes, she smiles just the right amount, and her delivery is both graceful and appropriate for a teenager. She utters an occasional nervous giggle but never strays from her talk. She's eminently likable. She later says that she doesn't memorize her speeches because it tends to make her more nervous. She prefers to improvise.

Near the end of her talk, Olivia gives Andy (who's here with Tommasina) a warm shout-out. While she shows an image of the two of them working in the lab together, she says, "Mr. Andrew Bramante single-handedly leads the research program at my school, and while its legacy speaks for itself, and has international winners every year, I think what speaks even larger is the amount of students that come back, even after they've graduated college, with a sandwich and have a conversation with him. And so I think it's really just crucial that we have people like him in our lives." To this comment, the audience applauds.

In conclusion, she says, "What I've really seen growing up in this time is that we're all so eager to accept the benefits that come from globalization and interconnectedness, and I think if we're able to accept these benefits, we also have the obligation to assume the responsibilities that come with this and now more than ever, nothing exists in isolation. There's no more 'over there.'

"We have to realize that we're only as strong as the weakest country among us. I think this is really underscored by a quote from John F. Kennedy, which is, 'Let us accept our own responsibility for our future.'"

More applause, loud and generous. Olivia's speech is without a doubt one of the most grounded and palatable talks of the conference. It lacks pretension and doesn't strain for effect. Afterward, she finds her parents and is exhilarated from the high of having delivered such a stellar performance. She beams and says it's so nice to have people she knows in the audience.

It's an especially impressive feat given that two days earlier, she received the pulverizing news that she did not receive early admission to Harvard. Like William, she was deferred.

Standing in the Guggenheim, her mother, Julia, says, "I'd be lying if I said she wasn't upset." There were tears followed by self-doubt.

"Look, it's a broken system," says Julia, stoically. "She only has A's on her transcript. She's never had a B."

Later, after Olivia had some time to digest the news, she shifted from wounded to angry.

Similarly to William, Olivia's deferral begs the question, if a straight-A champion female swimmer who took the world by storm with her rapid Ebola test, and who also volunteers teaching special-needs kids how to swim, isn't Harvard-worthy, who is?

What Olivia seems to learn from the deferral is that she's not immune to feeling pained and embarrassed by the whole ordeal. Which hasn't appeared to be her take throughout the college admissions craziness.

A few weeks earlier, she was sitting in the lab at a table with some other seniors. Margaret Cirino, a fellow four-year veteran of Andy's class, was looking at her detailed spreadsheet of her college applications. Margaret, who can best be described as extremely poised and self-reliant, has become openly terrified, neurotic, and stressed about college. Her older sister attends Stanford. She feels like her parents have essentially sacrificed their personal preferences and happiness to live in Greenwich so their girls could get a five-star public education. And now here she is, at what feels like the defining moment.

Margaret was lamenting the shame and embarrassment that will befall her if she's rejected by her top choices.

Olivia stood up, appearing mildly irritated by Margaret's agonizing, put on her backpack, and said with an air of slight disgust, "Ugh. Why do you care so much what people think? It doesn't matter." And with that, she walked off. Her statement and tone seemed to convey a sense of immunity from public opinion.

But when the Harvard hatchet falls, Olivia's not remotely cavalier about it. In fact, she's refreshingly, humanly candid about the impact it had.

"It was really hard for me and made me hate the whole college process," she says. "It's disgusting how people will give so much money to get their kids into a school. You could save a small country. I was so mad.

"I thought, what did I do wrong?" she adds. "I thought people would see something wrong with me, like, she didn't have high ACTs. Or, after Google, she stopped trying."

Despite being showered with public accolades and launched to superstardom, she's also been met with critics and skepticism, some of it catty and mean.

"Oh, there are definitely the haters," she says. Following her appearance on *The Late Show with Stephen Colbert,* she says people accused her of flirting with the host.

"I was just so nervous and I laugh a lot when I get nervous," she says.

Olivia has reflected upon the possible reasons she's become a target for some. "I think it's because I'm not a stressed person, it might come off to people as effortless or not trying or luck or unfair," she says. "Some people have a sense that I don't deserve what I've gotten." The online haters have sniffed that her success was owed to being from wealthy Greenwich and well-off parents—low-hanging fruit as insults go, but hurtful nonetheless.

She's also experienced the downside of fame. Though she's enjoyed many of the cool things she's gotten to do, such as going to the latest *Ghostbusters* premiere in L.A., she has gotten irritated when people have recognized her in public. On the train to New York City headed to a music festival, some guy loudly pointed her out—which she didn't appreciate.

But she knows this is all somewhat of a self-fulfilling cycle. The more media she does, the more speeches she gives, the more events she attends—her star is not going to fade. And these are choices she's making. But who can blame her? What high school senior doesn't want to walk the red carpet at a movie premiere or wear a princess gown for a night in Paris?

"For me, it's hard being in this limbo. I kind of want to be doing it when I want to be doing it and then not at all," she says. "I take responsibility for how public I've been after Google."

What's disturbing and disheartening is that, as an accomplished young woman, Olivia seems to have been subjected to an especially eager, voracious flock of vultures who delight in gnashing her up and spitting her out. Would a boy be accused of flirting with a female talk show host if he nervously laughed on air? Would people feel the need to say that a boy's success is attributable to his affluent zip code?

Olivia has been bestowed with a dizzying amount of glory and attention for her work, but that ever-present group of detractors who feel the need to stick her in a box they deem nonthreatening has made it hard for her to fully enjoy it. It's made her wary and cautious. She has the sense that some of the hatred stems from the fact she's well-rounded and not a nerd.

"I think there's this weird sense that if you are smart as a girl, you have to be only smart," she says. "You can't care about appearance. A boy can be smart and a partier. As a girl, you have to be in your room studying."

While some have looked for ways to snuff out her glory, Olivia has consistently turned to the most empirical and irrefutable thing in her life: competitive swimming. Why?

Because despite the fact that she won Google and other competitions, people could nitpick and do their own postgame analysis to undercut her achievement.

"I struggle with the subjectivity," she says.

Not so with swimming. "It's on the clock, you can't argue about it," she says. "It's a clear-cut numerical. It's easier to prove yourself." She also took solace in the fact that her swim coaches, unlike many others, treated her exactly the same despite the halo of fame encircling her. They saw her value in the pool and that was good enough for them.

Early acceptance to Harvard, in her mind, could have undercut her naysayers, shut them down. (It would also have meant she could swim in college, which she's always wanted to do. She's accomplished enough to swim at many Division I schools.)

"I wanted that validation," she says. And when it doesn't come, for a brief time, her starry world falls off its axis.

23

Danny

When human contact at the castle has been made, you hear the steady bark of what Leah Slate cautions is her attack dog. Once you make your way through the spooky room on the ground floor, which functions as a staging area for deliveries, mostly cases of fine wine, and up a staircase, you meet Montauk, a calf-height Havanese with an underbite.

Leah leads guests to a sitting room that's part of the original house, built in 1905, which is different from the "new" part of the house, the expansion added in 1916, which features a room that looks and feels like a salon where matters of great import are discussed. There are thick stately drapes that cascade from ceiling to floor, sculptures and armchairs and views out to a large reflecting pool. Off the salon is a wing that houses Richard's den/workspace. It's filled with, among other things, guitars that he collects. To show every room of the castle would take hours, so in giving the abridged tour, Leah sort of traipses through whole dark sections that are rarely used.

Back in the original section of the castle, with its dark mahogany moldings, Tiffany lamps, and jacquard-covered sofas, the sitting

room is a puzzle piece of a space that quickly leads off to more hallways and nooks. The room gives off a sense of mystery because its
asymmetry and placement give you the smallest of glimpses around
the rest of the castle, and how can you not wonder what's around the
many corners? As a child, Danny regularly got lost in his own home.
If you were to play hide-and-seek here, you'd have to draw space parameters or children would vanish. And forget about playing on the
castle grounds at large.

Leah is a petite, young-looking forty-six-year-old who's wearing
dark jeans, a camel-colored sweater, and flats. She works part-time as
a contractor for a management consulting firm and is very involved
in Greenwich town politics. She tucks her legs beneath her as she
talks. While her casual and comfortable appearance can't quite offset
the noncasual castle backdrop, her manner does. She's an easy smiler
with a quick wit who banters readily. In a nod to the noticeably di-
minutive stature of all the Slates, she says, "We didn't breed my children for height or athletic ability." She notes that Danny's lack of
athleticism was obvious when he was a toddler. "Constantly I felt like
I had to be saving his life. He'd be sitting on a chair and then he'd be
on the floor and I'd be like, how did that happen? You were just sitting
on the chair, why did you fall on the floor?"

One of the hallmark, defining dynamics of the Slate family children is their complete and utter inability to do an activity together
unless it involves winners and losers. They must vie for everything.
She cites the famous cooking contests. "Why can't my kids just cook?
No, it has to be a competition." Leah's the kind of mom who wouldn't
blink if her kids wiped their jelly-stained fingers on her blouse but
doesn't want to see anything less than an A on a report card because
she knows that with her offspring, less than an A is not an intellectual
shortcoming but the result of coasting. She knows this well because,
like her son, she was a quick study in all things related to school. She
says she and Danny can do little and get an A−. Richard and Michael,
by contrast, have a fixation with total mastery—they would never
stand for anything less than perfection and will practically recode

their own DNA with information. Leah and Danny would use that time to pick a new TV series to which they'll devote precious binge-ing hours. ("Sidney's jump ball," says Leah, characterizing their daughter and middle child as a hybrid.)

And then there's a whole other, rather shocking dimension to the Slate upbringing: their zeal for fast food. Here's Leah, living in a castle and thinking nothing of letting toddler Danny eat Wendy's and Pop-eyes and watch TV for hours on end. It's not out of recklessness—or even indulgence. "You can only be who you are as a parent and I'm not a very disciplined person, for better and certainly for worse." She says this candidly and with no attempts at false modesty, but with her Harvard education and Yale law degree, she certainly does not come across as someone lacking discipline—unless discipline is measured only by the ability to put down the clicker or drive by and not through Wendy's.

But in this town, a place where bringing your child to Popeyes is practically considered child endangerment, an Ivy League mom who lets her kids do things that People Like Her don't do feels exotic, liber-ating, and naughty. Yet when the Slates reveal these facets of their life, there's no sense of them trying to shock. They're matter-of-fact and slightly sentimental about certain aspects of it. Danny and Leah both speak with pride about his ability to order his chicken nuggets at the Wendy's counter at the age of two.

Despite her laid-back stance on junk food and television consump-tion, Leah is very clear that she wants her firstborn son to go to Har-vard.

"I do because I loved it," she says. "It was an incredible experience, I made my best friends, and I know for him that he would love it there."

For the Slates, Harvard has remained an important part of their lives. In fact, one of Richard's former law school professors, the at-torney Alan Dershowitz, attended Richard's fiftieth birthday party in Manhattan, where the family has a townhouse on the Upper East Side.

"We go back to every single reunion and we're so into it, it would make it harder if he didn't wind up there," says Leah, noting she'd be "sad" if Danny doesn't get in.

The family hired a college essay expert to help Danny, which is not unusual in Greenwich, where parents with means drop thousands of dollars for college admissions consultants and essay helpers. Danny has somewhat of a complex about his writing skills. He's actually a strong writer, but because it requires more labor than he's used to, plus revision, he's come to see it as a weakness, unlike, say, chess or math, where he can exert little effort and obtain superior results. (This is a theme with exceptionally bright kids who are used to learning and doing most everything with zero struggle; the minute they have to apply time and considerable effort, they perceive themselves to be deficient.)

Danny and Richard engage in ruthless speed chess matches. The elder Slate has zero interest in letting his son win. (This is the dad who jokingly called Danny a "pathetic worm" when his son got an inferior grade on a test.) The only edge Danny can get is to challenge his father in the evenings after Richard has had some wine.

Speaking of games, there's a competition of late that has ignited a fire within Danny. He walks into the lab one day in early December so excited, a fiery look in his blue eyes, it's like he's some superhero version of himself.

It's Assassin time.

The seniors at GHS, like seniors at many schools, engage in a ferocious game of Assassin. You get assigned someone to kill, and likewise, someone has been assigned to kill you. The art to the game is deducing who's out to off you and avoiding your assassin. There are also daily exemptions. For example, one day the exemption could be if you wear blue, you are immune from being killed. At GHS, the fatal deed is done by tapping your target with a wooden spoon.

Henry Dowling says, "By the end, it can be two people dueling each other with wooden spoons in the hallway. It's amazing."

Danny is in it to win it. And this is just the beta round. The real thing will happen in the spring. He's come in with his wooden spoon, gripping it like a sword. Takema walks in holding a curved dark wood salad spoon, one that looks far too fancy for the purposes of whacking someone.

Danny's pumped up. He says, breathlessly, "I know the person I'm trying to kill and I know who's trying to kill me. They tried to kill me earlier today and I outran her. Ran from music theory straight to Euro." It's hilarious and sort of restores your faith in the youth of America: even though there are countless videogames where kids can satiate their caveman urges to knock people off, those games can't compare with the visceral thrill of wielding a wooden spoon to assassinate someone in a high school hallway.

The kids try to one-up each other in all kinds of ways. William wrote some code that downloaded all the seniors' schedules so he and his cronies can strategically plot their moves. One kid sweet-talked Takema's schedule from a teacher.

Of these shortcuts, Danny says, "Fairness is for losers. People do whatever it takes to win by any means necessary."

Danny's Harvard essay is about whom he'd select to appear on his personal version of the *Sgt. Pepper's Lonely Hearts Club Band* album cover. It's a creative piece of writing that evokes his nerd charm and smarts. He chooses a smorgasbord of iconic figures, from Marie Curie to Gandhi to his siblings. He writes about how he got a record player for his birthday one year, along with the classic Beatles album. So right away, a reader sees he's the kind of kid who appreciates vinyl, who relishes the act of placing a needle on a record, a kid who delves into the Beatles. It's in keeping with his penchant for all things canonical. He reads the classics for recreation; it's unlikely he's ever laid a finger on a young adult book.

About a week before Harvard's notification day for early admission, Danny is sitting in the lab, his mushroom battery mentally and actually shelved, mulling over the somewhat mysterious and controversial Z-list.

The "Z-list" refers to kids Harvard admits a couple of months after all the regular decision acceptances go out, contingent upon a gap year. This is separate and distinct from the traditional waitlist. The rap on the Z-listers is they tend to be monied or influential people who fall short of Harvard's admissions standards.

Danny is reading an article titled "The Legend of the Z-List," published in the *Harvard Crimson*. From the April 3, 2014, article:

> When The Crimson first reported on the existence of a deferred admission program in 2002, 72 percent of the Z-listed students surveyed were the children of alumni. A follow-up story in 2010 confirmed that an exceptionally high proportion of students given deferred admission were legacies and nearly all came from affluent families.
>
> Was this secretive admission program a way to admit the children of ultra wealthy? A way to slip in unqualified students through the back door?

The online comments to the piece include several rebuttals, including from someone purporting to be a "non-legacy, nearly full-aid Z-lister" and people writing that the Z-listers aren't necessarily any less academically qualified.

Danny sits at one of the lab tables, reading this and considering what it would mean to him if he were to gain admission as a Z-lister. As a double legacy from a family who have donated money to their alma mater, he finds the whole thing distasteful. He doesn't like the fact that a Z-list admission would bring into question his merits and everything he's worked for. Plus, he'd enter college a year later than his peers, so everyone would know he'd gotten a slot by the skin of his teeth.

Despite his deep desire to attend Harvard, he says that if given the choice between straight admission to another prestigious school or having to enter Harvard feeling like a sloppy second . . . well, he's not sure what he'd do.

And so it is with a great deal of anxiety, fear, excitement, and anticipation that shortly before five P.M. on December 13, the Slate family is assembled in Danny's bedroom. Danny is at his computer, ready to log on to the Harvard admissions portal.

While they're biding time until five P.M. lands, the Slates make small talk, reminiscing about Danny's birth. Richard cried when his son came into the world but says he doesn't think he'll cry if his son gets into Harvard. Leah is outwardly nervous.

At five P.M., Danny checks his email and there's nothing from Harvard. And then, like William, who is doing this exact same thing across town, he logs in to the portal and starts metronomically refreshing the page.

All of a sudden, a screen appears that reads, "Admitted student form."

Wait, does this mean he's in? Where's the letter that says, "It's with great pleasure that we inform you . . ." or "Congratulations!"

Danny's brain goes into panic mode, trying to decipher this ostensibly promising message, and without a moment further of hesitation, he hits "Accept."

Michael and Sidney are screaming and whooping. Leah bursts into tears. Richard, as forecasted, does not.

But Leah is worried. Something about this doesn't feel right. Where is the standard-issue, clear-as-day declarative letter? Danny is overjoyed, but his elation is tempered somewhat by what he describes as "my mother spending the next fifteen minutes trying to prove I actually DIDN'T get in."

Danny says once he hit "Accept," that was good enough for him.

"I accepted the offer before they could take it away," he says, without irony.

Some of his friends who have known tonight was notification

night had warned that they'd storm the castle to celebrate Danny's admission, and they do. Richard opens a bottle of wine, Leah sends out an email to family. By 5:31 P.M., she has relaxed into the good news and sends a text that says, "Good news here. Over the moon excited. Hoping for good news for others."

24

Sophia

Why Sophia's Lyme samples shrank and hardened and were nowhere close to growing biofilm will always be one of those unknowable scientific mysteries. Without the benefit of a professionally sterilized lab, she has no choice but to grow her samples in a lab where kids eat greasy pizza, gooey nachos, and egg sandwiches. Not that the food is the cause, but it's just impossible to know what contaminated her first batch.

So she starts again. After her failed trial, she doesn't waste time before defrosting more rabbit blood and building her chemical layers around the Lyme bacteria inside the well plates. But she's decided to change one thing, the only thing she did differently from last year.

"I'm not putting the trays into the glove box," she says. She can't say for sure whether that environment was the culprit, but since it was the only new factor, she's not willing to give it a second chance to foil her. Plus, she doesn't have time. She caught the very tail end of last year's science competitions because of how complex her project became, but missed CSEF. This year, Sophia is determined to apply to as many of the major fairs as possible. The first one is the CT STEM Fair and the deadline to submit an application is December 21.

To keep kids on track and to give them a sense of where they stand, Andy tapes a spreadsheet on the front of one of the cabinets. It has the students' names, their project titles, and a brief description of what they're working on. He keeps it very simple, using basic color coding to signal everyone's status. If you're in good shape, your block on the spreadsheet is emerald green. Positive but not quite there yet is mint green. Yellow means you're in shaky territory. Pink is bad news, and red is "Get your shit together." When the status sheet goes up, the kids gather around and look at it, pondering their progress. Andy has found that the sheet, hung without any fanfare, has an immediate effect on everyone's efforts.

Andy's modus operandi is to reward effort, not merely outcome. So even though Sophia's project has a giant question mark hanging over it, as nothing can even be tested until the tried-and-true biofilm appears and proliferates, she's solidly in emerald green.

For Andy, who tends to view the kids through a series of lenses based on how much heavy lifting they require, how well they manage their time, and how easy or not they are to work with, Sophia's the ideal research student, an exemplar. She's a self-starter who needs no cattle-prodding and comes to him with smart, logical questions. Some kids require his reassurance at every step, checking in before they remove a beaker from a shelf or pour a chemical into a flask.

Sophia also takes her spot in the class seriously. After getting to know Andy, Gregg said to his daughter, "That guy's amazing. Teachers like that don't come around very often. You're lucky to experience a teacher like that."

On a cold December morning, once she finishes pipetting her concoction into the trays, Sophia places the well plates in an old incubator that lives in the smaller equipment-stocked room across the hall.

She tapes a note on the incubator warning people in big letters "DO NOT OPEN!!!"

While she waits to see whether the biofilm will grow, Sophia can begin to create the drug therapy, the mixture of antibiotics and PDE4 inhibitors that she'll administer to the biofilm to see whether they eradicate it.

One day she's in the lab dissolving the antibiotic tetracycline (in powder form) in water. It turns a bright yellow, emanating a sunshine-like glow.

In walks Romano.

"Romano," she says, acknowledging his presence and smiling.

"What's up, Sophia?"

Sophia and Romano are in the same friend group, the primero popular girl-boy posse twenty kids strong who often break off into smaller factions. These are the ones who sit in the unofficial but well-known popular zone in the cafeteria, who go to parties and play sports and benefit from having catalog-worthy good looks.

Romano and Sophia sometimes hang out together. He's even cooked Italian food for her at her home. Gregg was noticeably impressed when he came home one day and, unlike so many of the kids who come by, lounge on the furniture, help themselves to food, and barely acknowledge the parents, Romano jumped up, shook Gregg's hand, and engaged in conversation. A teenaged unicorn, alas.

A naturally shy, low-key person, Sophia seems to have been anointed popular, as opposed to some of the overt queen bees who clearly have anointed themselves. They strut through the cafeteria with their best resting bitch faces, telegraphing their social seniority. To anyone above the age of eighteen, it's comical. But then you remember that in the hive of high school social politics, the queen bees can unleash real venom unto the less powerful.

Sophia finds the inevitable drama and the group texts tiresome. Every decision about where to go and with whom gets endlessly batted around. There are angry flare-ups if the boys break rank and hang out with—gasp—sophomore girls. There are couples and breakups and tears and retaliation.

And it will surprise no one that popular is not synonymous with smart. Sophia and Romano are among the highest achieving of the popular kids. So serious is Sophia about her grades, she thinks nothing of studying on a Saturday night. But she's not so removed from the scene that she won't go to the occasional party.

Gregg and Stephanie take a laid-back parenting approach to the high school party scene. They trust Sophia and Brandon not to do stupid things and accept that they will go to parties. But their kids are mild in temperament and that translates to their social behavior. For instance, when Sophia categorically says, "I would never throw a party in my house," there's nothing about this statement that strikes you as negotiable. She also seems protective of her lovely, pristine, post–Hurricane Sandy home, her final abode after those two unsettling years of wandering around Greenwich.

For the next week, Sophia waits patiently on her well plates. She periodically checks them, hoping that the creepy biofilm will start to appear.

One day, Sophia peers into the incubator and sees what seems to be biofilm action. The smooth surface of the carefully measured chemical layers has begun to be taken over by a crusty top layer. Sophia smiles. This is what she's been waiting and hoping for. No one has ever been so pleased to see the arrival of crud.

Her next steps are to try to treat the biofilm with a series of permutations of three PDE4 inhibitors, the drugs that have been shown to remove biofilm that grows on devices implanted in the body, with two different antibiotics. She'll need to test each of the PDE4 inhibitors alone, the antibiotics alone, and then all the various combinations of the two kinds of drugs. Once her drugs are administered, she'll need to measure the biofilm thickness each day to see if the drugs are having any effect.

One day in the lab, she's got her well plates out on the table—four trays with twelve wells the size of a quarter, each containing biofilm. She's pasted a diagram/map of her plates into her lab notebook so she can keep track of what combinations she's administered. As Sophia works, she refers to her diagrams and carefully pipettes the drugs onto the biofilm. This is a time-sensitive situation, as the biofilm will deteriorate into an unusable form in about two weeks, so it's critical that Sophia works efficiently.

She's first administering the PDE4 inhibitors in combination with the antibiotics tetracycline or doxycycline. There are nearly one hundred PDE4 inhibitors. Sophia's chosen three to test on the Lyme biofilm: rolipram, roflumilast, and cilomilast. The first, rolipram, has been used to suppress tumors. Roflumilast and cilomilast have been used for inflammation, which is thought to be implicated in Lyme disease.

For the next several days, she checks on her trays and measures the biofilm thickness with a ruler.

After several days, she can see that the biofilm is beginning to break up. It appears that the drug combinations are working. Every day she comes to the lab, it seems there's a direct correlation between the degradation of the biofilm and the width of the smile on her face.

Sophia's never been a bold or supremely confident personality, despite her winning combination of beauty and brains. It's as if it's never occurred to her that she possesses either. She tends to fold her body inward when sitting, as if she's trying to make herself as small and discreet as possible. She's not boisterous and she never, ever brags. She speaks softly.

But when she works in the lab, there's a palpable, visible difference in the way she carries herself. She has presence; she stands taller. She moves carefully, but with no hesitation.

Her parents have noticed. "First of all, Andy has brought her out," says Stephanie. "He gets them to push themselves." When Stephanie took Sophia to the Norwalk Science Fair at the end of Sophia's sophomore year, she was stunned to see her little girl in a suit, about to expound on her complex science project.

"That's when I really started seeing her becoming a young woman and that she's really going to go somewhere," says Stephanie.

Because she had such a great experience in Andy's class last year, Sophia decided to do a summer biotechnology session at the University of Michigan. She went on her own and studied DNA analysis and coding. It was unlike any experience she'd ever had and she liked it, even if it was "nerdy."

Now that she's working through the various drug combinations, it looks as if the rolipram in combination with either tetracycline or doxycycline is the best for chiseling apart the biofilm.

Sophia also tests the PDE4 inhibitors alone, and they appear to have some effect, though not as potent as when used with antibiotics.

Finally, it's all coming together. There's just one issue, and it's a big one. Neither the eyeball measure, photos, or even a ruler are empirical enough to actually demonstrate or explain that the drug therapy is working. She's going to need to figure out some metric that can definitively show that the drugs actually degrade the biofilm.

Alas, one bit of incremental success yields a new quandary. But for now, Sophia's taking what appears to be a win.

25

Andy

"**A**re you fucking crazy?"

That's what Kenny Uliano asked Andy upon hearing that his friend and colleague at PerkinElmer was planning on leaving the company to teach high school.

"Jump back twelve years; his daughter was only little. He had a good career. Everybody knew his qualifications," says Kenny, a PerkinElmer lifer who's spent forty-four of his sixty-five years with the company. "He could have been in five different parts of the company, but he chose not to."

Kenny and Andy met in the technical support side of PE, a company that designs and manufactures scientific instruments. PE is notable for building, among other things, the Hubble Space Telescope mirror. Kenny's specialty is in thermal and elemental instruments, which measure heat characteristics and elemental composition.

Although Andy's work was with light spectroscopy, the use of light for scientific analysis, he and Kenny became fast friends, drawn by their slightly stupid, slapstick sense of humor and predilection for adolescent pranks.

Andy's career didn't start on the instrument side. His first real scientific job during the year between undergrad and graduate school was working as a flavor chemist—creating flavoring for everything from peach schnapps to cheese. He remembers the day a steel cauldron of brewing cheese flavoring exploded on him.

"I got on the subway and cleared out the entire car," he says with that certain male pride for grossing out the masses. "Boom, just like that."

It was a quirky job that made for interesting chitchat, but he learned that year that he wasn't gunning to become the world's most renowned flavor chemist. He also moonlighted, teaching chemistry at night school, first at Baruch College and later at Sacred Heart. He enjoyed it and it was easy money.

After he finished his master's degree in chemistry, Andy got a job with Foxboro, also an instruments manufacturer, as an applications chemist. He was running experiments on instruments used in industrial hygiene, which look at, say, the air quality of a workplace. The job made his tinkering gene do cartwheels; he loved working with the gadgets and gizmos. "It was like playing in a friggin' toy box," he says.

It was also the Foxboro job that took him, for the first time in his life, in his late twenties, out of the Bronx. He moved to Stamford, Connecticut, and worked in South Norwalk. But when Foxboro decided to transfer its Connecticut operations up to Boston, Andy took a rather parochial view of the situation.

"The thought of moving to Boston was like moving to a foreign country," he says. Looking back, he knows that was shortsighted, but he was happily dating Tommasina Cantone and wanted to stay put. He'd met Tommasina, a fellow Bronx native and Fordham chemistry major, at a chemistry party in college.

Andy worked at Foxboro until they shuttered the Connecticut operation. He then took a job with Hitachi providing technical support to sales teams trying to sell nuclear magnetic resonance imaging instruments, which are used for chemical analysis via magnetic rotations of hydrogen and carbon.

Andy was paid to fly around the United States helping sales guys close seven-figure deals. It was a sidekick-savant role. As salesmen would bloviate, grease palms, and make promises about their cutting-edge scientific instruments, they needed to be shored up by a guy like Andy, a design scientist who could explain, demonstrate, and legitimize the sales guys' claims without outshining them or making them look weak. And it didn't require him to do the screw-turning of a salesman. He didn't get paid the hefty commissions of his sales counterparts, but he was liberated from the smarm. Once a client purchased a machine, they'd get a training session with Andy, who would then take them out to dinner. "I was always the hero," he says.

So he was the traveling salesman's secret weapon, which meant he traversed the entire country. He says it's easier to talk about it in terms of places he *didn't* visit. "Never got to the Dakotas or Nebraska," he recalls.

The end of his love for the job didn't come in the form of a harsh or sudden breakup. It was more like a slow but steady power outage. Andy Bramante tired of the Andy Bramante road show. There are only so many hotel banquet rooms with bad carpeting and guys drinking watered-down vodka tonics that a man can take, and there wasn't much excitement to be found in the twentieth trip to Dubuque.

Plus, he was now married to Tommasina. They wanted to start a family, and he wanted to stay put.

Through a series of contacts, he got an interview at PerkinElmer, where he would spend the last decade of his career in industry. They put him in the fluorescence group, which meant he had to learn polarimeters and luminescence spectrometers, both used for chemical analysis. His job was trying to come up with turnkey applications that would make the instruments more salable.

Eventually he transitioned into a support role, assisting Perkin-Elmer's engineers and their clients. He was able to work a day or two a week from home, which was useful because he and Tommasina were now parents to baby Sofia. The gig had its conveniences, but it was hardly the job of his dreams. Plus, he had a boss who he sensed had it out for him. Seeing the pace Andy keeps during the school year, and how that metabolism is his natural state, it's easy to imagine that working from home fielding phone calls would drive him bonkers. Meanwhile he continued his adjunct teaching gig.

Kenny Uliano marveled at his friend's moonlighting. "When do you have time?" Kenny would ask him. "He taught, he did all the support stuff. He was a brand-new father. But he always had this energy that was never-ending. He always made it look so easy."

In the late 1990s, PerkinElmer started thinning out its ranks. The family-owned feel that pervaded the company when Andy arrived was no more. And then there was that boss. Every year at his annual review, Andy would sit down with the guy beneath that soul-destroying corporate fluorescent lighting and the guy would say, "Andy, I don't know what we're going to do with you." Because of his seniority and prior roles, he was making significantly more than his peers in the same job, something Andy sensed rubbed his boss the wrong way.

Andy felt the end was nigh. He didn't know how or when he'd get driven off the road, but he sensed it was better to start looking for an exit ramp on his own.

And then one day in 2002, Andy was at a meeting for the American Chemical Society where some teachers, one from Greenwich High, were giving a talk about high school science research programs. Andy stopped in with the thought of maybe donating some old instruments to their labs. But as he heard about the research programs, something clicked.

This was it. This was what he wanted to do. He loved the thought of working with kids on truly innovative ideas and shepherding them through these competitions. There were appealing stakes involved that would provide both framework and goal.

And there was something, or rather someone, who had been encouraging him for years to consider teaching high school. None other than Vinnie Bucci, Andy's high school pal, one of the ushers in his wedding, a lifelong friend.

The Booch.

Vinnie, Andy, and Arpad Dobransky formed a geeky triad in the Class of 1980 at Mount Saint Michael Academy, an all-boys Catholic school in the Bronx's Wakefield section. They were in the smart kid classes and gravitated toward one another in the way that unformed nerds who later prevail in life tend to do.

Vinnie went to Boston College and then straight to Brown University to get a master's degree in education. He's been a public high school English teacher his entire career. (Arpad went into finance and moved to Jacksonville, Florida. The three are still in touch.)

Standing in Andy and Tommasina's house today, their yearbook from "The Mount" spread open on the counter, Vinnie deadpans, "I know when you look at us back then, it's actually impossible to imagine that the women weren't all over us."

Teaching has been Vinnie's life's work. He loves it. And it was Vinnie who was the first person to sense that his old pal, Andy Bramante, might be a gifted teacher.

Andy even fit the bill from a casting perspective: short, small framed, unintimidating. He has an easy smile and a natural air of trustworthiness. There's a warm, everyman demeanor. No airs, no pretenses. Midconversation, he'll remove his glasses to wipe off any detritus, appearing not to notice he's doing this. If he ever had a bust-'em-up Bronx accent, it's been sanitized, but for one relic, like an archaic subway token that turns up in your junk drawer. He pronounces "huge" as "YUGE," much like two other native New Yorkers—Bernie Sanders and Donald Trump—who have been widely noted for being unable to flush it from their use.

One of his natural reflexes is to lend a hand, help people out—and whenever possible, save you a buck or two in the process. If in casual conversation you mention any niggling issue in your life, from a backache to your dog's ailing gallbladder to a leaky valve on your fifteen-year-old car, Andy immediately starts thinking of solutions, people he knows who can fix the problem ("I know a guy . . ."), websites where you can avoid retail pricing, you name it. It's in his DNA:

Figure shit out, do for yourself, don't get scammed, and don't pay people unless all else fails. This instinct to help, and to do so in a crafty manner, means he's both done and been done a lifetime of favors. This is a man who loans his auto mechanic a lawn mower so the guy doesn't have to buy one.

Every fall, Vinnie and his wife, Kathy, a pediatric oncology nurse, host a party they dub "Back to Normalcy." It's a nod to the fact that for Vinnie and his colleagues in education, the annual clock is reset in September, which is when they also regain their equilibrium. Andy and Tommasina always travel up to Salem, Massachusetts, for the party. For years, Andy had been listening to stories from Vinnie's teaching friends and colleagues about what they do, and in many cases, the stories culminated in some tale of making a difference in a kid's life. These accounts seeped into Andy's brain and stood in sharp contrast to his working life. He had had a good run, sure, but there was nothing in the way of changing anyone's life.

So when Andy sat in on that talk about high school science research, knowing his corporate life was looking and feeling bleak, nothing but a dead end staring back, he started to put a plan in motion.

26

William

William's universe, usually a vast and jumbled place where he steps over piles to hop to other piles, becomes extremely insular and focused after his Harvard deferral. In some cases, like with his scientific research, he's a risk-taker. But not so when it comes to college admissions. He's decided to apply to twenty-five schools—all eight Ivies, MIT, UCLA, UC Berkeley, Caltech, Carnegie Mellon, Duke, McGill, Yale-NUS (Yale's Singapore campus), University of North Carolina at Chapel Hill, Amherst, Tufts, University of Chicago, Stanford, University of Connecticut, University of Michigan, Northwestern, and Johns Hopkins.

Most kids apply to college using the Common Application, a sort of McApplication, and write one major essay seen by all the schools to which they apply. They then write supplemental essays for individual schools that are sometimes optional and sometimes mandatory. William had made the decision not to submit any supplemental essays to Harvard (based on advice from people in the GHS guidance department, who took their cues from Harvard's regional admissions person). Still feeling the sting of rejection (okay, deferral), he decides to write them for the twenty-four other schools—bringing his total essay count to fifty-five.

With all of the regular decision applications due in early January, William's holiday break becomes a kind of grim, monastic cave that feels almost punitive. He derives no joy from the personal-essay-writing conveyor belt. "I'm not a very self-reflective person," he says, which, of course, betrays a certain amount of self-reflection. Peeling back the corner of his soul doesn't come naturally to him. Add to that the fact that right now his soul is a very publicly wounded battlefield and the whole thing just feels miserable.

The night he got the Harvard news, he plodded ahead and went to the band concert. When he arrived, he was besieged with questions from his peers about Harvard. He became his own personal bad news ticker and had to watch the confusion and condolences wash over everyone's faces. He was cool about it, betraying no extreme emotion.

"I went home, went to bed, and it was fine," he says, adding that he mostly felt nothing at all.

But the next day, the wall started to crumble. William bumped into Mr. Yoon, the band director and controversial GHS teacher who was suspended in 2015 following a case of alleged verbal abuse toward a student and then reinstated by Greenwich's frighteningly powerful board of education.

Needless to say, Yoon is not known for being particularly warm and fuzzy. William has always been in the pro-Yoon camp, admitting that his band director could be unsparing but feeling overall that he was a very effective teacher. William hadn't shared his Harvard news with Yoon, but interest in William's college fate was such that he hadn't needed to. It had made its way to the band director. When William encountered Yoon, the teacher delicately put a hand on William's shoulder and said that "college admissions aren't fair for us Asian Americans."

Says William, "That was the first time it sort of struck me that Mr. Yoon, my band teacher, cares for me. It was really strange."

What this gesture—gentle, minimal, and unforced—did was allow for a relatively untapped emotion to set in for William: vulnerability. The quiet acknowledgment that he'd been dealt a super crappy blow and had every right to feel its pain softened him. It was perfect in that it lacked pity, which would have had a curdling effect on William.

He says, "I remember thanking him and I was sort of starting to break down and I sort of walked away. I went to the bathroom and calmed myself down for ten minutes and then went about my day."

Despite his natural instinct to move on and remain task focused, William can't help but do a fair amount of self-reflection and explanation seeking as to why Harvard passed on admitting him early. But there's no satisfying or objective answer. The only other Greenwich High student admitted early, besides Danny, is fellow research student Henry Dowling. Like Danny, Henry is a legacy: his father attended Harvard. Like Danny, Henry is a bright, multifaceted high achiever. Their family situations likely helped them get in, but *not* being a legacy doesn't explain why William was left out.

The GHS guidance department knows the kids can be cutthroat and that the spirit of competition among the academically dominant can often veer into the unhealthy. What's less clear is whether the department's approach helps or feeds this mentality. Example: kids are told not to tell people where they're applying, both to spare themselves any embarrassment should they be rejected and also to keep others who may get rejected from feeling bad. Transparency is actively discouraged.

The problem with this kind of mollycoddling is that it implies that college acceptance *is* the supreme goal above all else and that to get rejected is absolutely humiliating. Likewise, to get accepted at the expense of someone else is to wound him or her. So the best thing to do is skulk around in the cloak of secrecy, keep your college spreadsheet to yourself, and neither rejoice nor mourn.

But since this is high school, the secrecy just raises the speculation and intrigue and places everyone on edge. (Duh.) Because of his ultra-high profile, William's Harvard deferral very much becomes a public affair, a kind of cautionary tale/cause célèbre. Kids speak to him with a funereal tone, and there are certainly some people, his academic foes, who revel in his stumble but have the smarts not to show it.

L'affaire du Harvard has had another effect on William, which is that he's getting more intent on formulating a science research project. And it's a doozy, more sophisticated and far-reaching than anything he's done in the past. After spending hours researching and reading papers, William has mapped out what he thinks is a blueprint for a diagnostic test for Alzheimer's disease, the form of dementia charac-terized by plaques and tangles in the brain that rob people of memo-ries and function. It's almost always diagnosed based on symptoms. At this point, an empirical diagnosis can only be made at autopsy or through a much less commonly used PET scan of the brain in combi-nation with a radioactive dye. The latter isn't typically covered by in-surance and so has not come into widespread use.

Alzheimer's isn't the only type of dementia and is sometimes con-flated with other forms. The reason it's so hard to see if the brain has been invaded by the telltale plaques and tangles is because diagnostic chemicals can't easily cross the blood-brain barrier, the network of vessels and tissue that separates blood from brain fluid.

A cheap, fast way to diagnose Alzheimer's disease is one of the holy grails of modern medicine. And now seventeen-year-old Wil-liam Yin thinks he has a potential fix to this problem—one that has eluded scientists for decades.

This is precisely what Andy means when he talks about how young minds are often ripe for scientific innovation. Yet he has very mixed feelings about William's sudden drive to do a project. For one, while he knows that William loves research and is truly gifted in matters of science, what's now motivating him is his urgent desire to show Har-vard something they'll think is admission-worthy. Andy doesn't believe in the tap dance, or doing any kind of veiled grovel, especially for someone like William. It's not unlike the way Andy felt asking for a $4,000 raise. His take is simple: William should have gotten in, he didn't, Harvard's loss and stupidity, the world has plenty of other excellent schools. Go where you're wanted, don't beg or try to woo, it's not worth your time. Especially when you have those fifty-five essays to write.

It all sounds so logical. But to a kid whose family has rationally or irrationally had their dreams pinned on this one school, he can't and won't take the "no," or the "not now," easily. William holds out some hope that he can prove himself to Harvard.

In early December, prior to the Harvard deferral, he walked into the lab after school with a broad conceptual plan for his Alzheimer's diagnostic, a mind-boggling, highly theoretical road map. It had imperfections, which he was quick to point out. He was questioning his patchwork plan even as he described it, debating whether it was worth the pursuit.

But after the Harvard snub, the ambivalence is gone. He's going for it.

27

Danny

On December 14, both Danny and Henry Dowling walk into the lab and before uttering a word about their Harvard acceptances, say, "Yin should have gotten in" and "Yin deserved this." There's a complete absence of gloating or showmanship. They also seem to contain their own happiness for fear of seeming churlish. They don't go so far as to say that their legacy status likely had a hand in their acceptances, but other kids do. It's not even mean; it's said with shoulder-shrug resignation, like, "Duh, obviously that helps." Danny and Henry hold firm status in the smart kid set, so no one says that they shouldn't have gotten in; the shock and unfairness is that Yin didn't.

Danny may owe some of his Harvard outcome to Latin, that deadest of languages. For his alumnus interview, he caught a lucky break when he was randomly matched to an ostensible adult version of himself. When Richard dropped off his son at the alum's house, he expected Danny to walk out forty-five minutes later. The interview lasted two hours.

Danny and the alum are both Latinists. The Harvard guy is a Chicago Cubs fan, and he and Danny discussed the team's recent World Series victory.

"He brought up how when he would go to the stadium, there was a sign that said, '*Eamus Catuli*,'" says Danny. His eyes dance when he recounts this because in that instant, the conversation landed right in one of Danny's sweet spots.

"*Eamus Catuli* translates to 'Let us go, Cubs' instead of 'Let's go, Cubs.' What I said was it should be '*Ite Catuli*,' which would be 'Go, Cubs.' '*Eamus*' is a subjunctive and '*ite*' is an imperative. There was a mistake."

Danny even pointed out that another translation for "*Eamus Catuli*" is "May we go, Cubs?"

In other words, he smacked a leadoff homer in the top of the first. Danny's face gets all buttery and warm just thinking about how he so nailed it. Seriously, what are the odds that you'll open a conversation with a total stranger about the World Series and then make the completely illogical turn to Latin grammatical moods where you can illustrate your nuanced expertise? He had the guy at "*Ite Catuli*."

From there, the two had the most fluid of conversations about everything from science to music to Danny's food allergies. Danny is a stellar conversationalist, with the intellect to hold his ground on just about any topic. And his affable demeanor keeps him from ever seeming pretentious. He gets nervous in crowds or large parties where he doesn't know people. But in a one-on-one setting, he shines.

There are some kids who do better with adults than with their own peers and Danny is probably one of them. He knows a little about a lot—hence his knack for crossword puzzles—and like his mother has a quick wit. If the art of bantering were offered as a course, he'd get an A+.

But despite the fact that he's now been accepted early to Harvard, Danny is far from relaxed. On the contrary, he's constructed a new tent of pressure that he's inhabiting, one that he describes as the "fear of fucking up and that they'll take it away from me." Because he's so bright and can get away with doing very little, he knows he can tunnel into a black hole of spending multiple nights watching reruns of *Curb Your Enthusiasm*. And his best procrastination excuse—focusing on college applications—just disappeared.

In regard to his mushroom battery, Danny underwent a power outage with the project the last couple of months while he was working on his Harvard application. The Regeneron deadline came and went. Danny says he was a little disappointed that he didn't make it, but "I'm going to do CSEF, I can still do CSEF." He plans to restart his engine and focus on the state fair, which he's never been to.

Andy notices the missed deadline and is wary of Danny's overtures about making CSEF. He's seen this movie before. He's not completely giving up on Pizza Man, but he's not counting on things to happen.

What seems to be the case is that the class is a mental and social oasis for Danny. It's become a realm for him to decompress from all the less forgiving and more structured aspects of high school. There can be significant downtime in the class—while kids wait for materials to arrive or for their experiments to bear out. It's not unusual for the kids to use some portion of science research class to do homework, study, or even—and this one irks Andy to no end—watch YouTube videos on their Chromebooks. Andy's relaxed about most of this because, given his load, there's always someone working, something that needs doing, some fire that needs extinguishing.

So he'd take no umbrage with Danny's tendencies if Danny would just put in the bare minimum. Even the kids mock his procrastination. Which is all part of Danny's standing in the class, a place where he's arguably at his most social. Outside, he's not a voraciously social guy. He doesn't throw parties at the castle or frequently play host. In fact, the castle seems at times to have placed some distance between him and the non–castle dwellers, i.e., everyone else. Kids have ribbed him for living in such a place (as if any kid has a say in the family home) and sometimes make cracks about his family's wealth, pointing out that nothing monetary is a hardship for Danny.

When asked whether he's embarrassed about his abode, he says,

"When I was younger, like, in middle school, I was. But no, not now. My dad told me I shouldn't be embarrassed and I'm not."

Plus, a large public school like GHS can serve as somewhat of an equalizer. Here in science research he's got a good deal of social currency from being Pizza Man, and the other kids seem to regard him as a curious hybrid of worldly sage and daffy court jester. They smirk and chuckle when they say his name because he spends so much time not doing research, but it's with an enduring affection. (They also love pulling the occasional prank on Danny, such as the day this fall when, for no apparent reason, someone zip-tied the straps of his backpack together. He suspected it was Madeleine and Derek. While Madeleine readily admits this is something she *would* do, she says she wasn't the perpetrator this time.) Watching the social dynamics swirl around him, you get the feeling nearly everyone in the room would post bail for him if he ever needed it. Which, of course, he wouldn't.

So, how is it that you come to live in a castle?

Given the family lineage, it's not that surprising. Three of Danny's grandparents have Ph.D.s. Leah says her father was a bit of a prodigy, graduating from the Bronx High School of Science in three years and eventually earning a doctorate in operations research. He was always into technology and found significant success in the world of venture capital, where he's considered a luminary, having invested in dozens of companies, including LinkedIn, Blue Apron, and Uber.

Richard's scientist father worked for Bell Laboratories and his mother taught at the university level before turning to fine art painting later in life.

After graduating from Harvard and then Harvard Law, Richard went into finance. At the time he and Leah married, Richard briefly worked for the notorious hedge fund SAC Capital, the target of multiple SEC investigations in the early 2000s, but left and eventually started his own fund, which did extremely well. Danny has a deep admiration for Richard's intellect and ability to seemingly master all of his pursuits, which range from chess to guitar. He says simply, "My dad is good at everything he does."

But the stress of managing considerable sums of other people's money became too much. Leah says, "He sort of said, 'I don't enjoy this business that much and I'm not that greedy.' So he was at a point where he felt it was good and he retired."

This happened when Danny was in elementary school and Richard was forty years old.

As for Leah, she grew up comfortably and went to a private all-girls school outside of Philly. While her upbringing didn't resemble that of her children, she never wanted for anything. Her family had a nice house and took nice vacations.

She says Richard, by contrast, "always had a sense of not being well-off as a child." Add that to the fact that he got what Leah calls the collector's gene, and one could argue the castle is really just a storage facility for his many acquisitions. Leah believes they possess the world's largest collection of antique Steiff stuffed animals and puppets. (Steiff is a German manufacturer of high-end plush toys.) A new shipment recently arrived and that lot are spread over the dining room table.

Richard also collects wine, guitars, and Dutch art. There's a building on the property that houses the overflow. Leah jokes, "My house is like high-rent *Hoarders* now. He buys everything."

Given the fact that Greenwich has an assortment of private schools and the Slates have the means, it's interesting that they chose to send all of their kids to public school. Leah says she looked into both options, but she and Richard both intuitively knew they wanted public education for their kids. Richard went to public school and had a good experience. They were both drawn to Greenwich's gifted program, the Advanced Learning Program (ALP), which starts in elementary school.

And socially, they liked the ethos of public school. Says Leah, "That's exactly the ethic we want our kids to have, which is you have to work hard to make your way in the world. Whether there will be cushioning for you in other things because you're fortunate to be from a family of means, you need to achieve—that's where your self-esteem and happiness is going to come from. And it's also the only thing you can really rely on."

28

Romano

Now that Romano and Andy are settled on gelatin as the base for his liquid bandage, Romano is beginning to assemble his concoction. He's both excited and tentative about finally starting to experiment.

Andy's reading the box of gelatin, that neat little pocket square of a box that Romano bought at the grocery store.

"I've never made gelatin before," Romano says. Andy's not concerned, since following the instructions on the box is the least of their worries.

"I know this molecule now, I get this molecule, and for this to really be my own thing, I want to do this," says Romano, psyching himself up, like a boxer about to step into the ring. "Forget about the science fairs and the judging and all that, just for my own science research, I want to make it my own."

Andy pulls out a measuring cup and some beakers.

Romano divides the gelatin powder into two small glass beakers, which are placed on a circular heating element. He pours in some room temperature water. Andy hands him a pair of safety goggles and a glass stirrer and watches as Romano gently mixes the liquid. A gooey slime quickly forms—and makes its presence known in more ways than one.

"It's nasty," says Andy, scrunching up his nose.

"It's already gelatizing," says Romano. It remains to be seen whether he'll successfully produce a liquid bandage, but he's manufacturing new words at the very least. *Gelatizing.*

"It's like one big booger," says Andy.

"It's going to be called the liquid booger Band-Aid," says Romano, laughing.

Romano stirs the boogery glob at almost slow-motion pace.

"Have you ever been in a kitchen?" says Andy. "Come on, man, stir it. Don't make love to it." Romano kicks it up a notch and starts rapidly swirling the snotty mixture.

"There ya go," says Andy.

Andy steps away and moves over to Verna. As Romano stirs, he registers the noxious, cruddy odor wafting up from the beaker. His face suddenly looks like a bunched-up dishrag.

"Oh, my god," he says, wincing. "This smells like an old Italian man's foot." The kids in the lab cut a wide circle around Romano and his brewing stink.

Once he has the two beakers of gelatin, Romano measures out a quarter gram of L-DOPA, a fine white powder, into a little plastic trough. He pours it into one of the beakers, using the other as a control. He wants to see if gelatin plus L-DOPA is stickier—that is, whether the L-DOPA will impart its adhesive properties to the gelatin. One, uh, sticking point is that with the infrared radiation spectrometer still broken, there's no way for Romano to take a chemical footprint of the mixture, to know whether the L-DOPA is, in fact, taking hold in the gelatin. That's going to have to wait.

Is the time going to cost him? He'll find out.

The start of December isn't fantastic. Much to Romano's disbelief, Team Caitlin is still unleashing its fury over the breakup. A bunch of Caitlin's friends made and blast-texted ten-second videos of them all saying, "Fuck Romano!"

Really??? thinks Romano.

"We were, like, best friends and now we're like enemies. She was the most genuine, nicest, sweetest girl. Maybe I caused it by breaking up. But I've never seen this side."

Romano finds the whole thing upsetting and irritating, especially because the entire time he was with Caitlin, he felt there were so many people rooting against them as a couple, desperately wanting to see them fail. And even though they dated for eight months, the naysayers prevailed.

There's an aspect to his life that helps keep him from getting stuck in the tar pit of high school drama. Romano is a committed Christian. Although he was christened Catholic, a few years ago, at the suggestion of his mother, he and his older siblings joined a Christian youth group called FOCUS. Kim's not particularly religious, but she wanted her kids to have a spiritual compass, a more structured place to think about being a good, ethical person.

All the Orlando kids embraced FOCUS and say they get a ton out of it. There's nothing evangelical about them. They don't proselytize or even talk about anything Jesus related. But Romano says it's been hugely beneficial to him and keeps his outlook balanced. He's drawn to the overall gestalt of the Bible—be a good person, treat people well, and put in a little time toward not being an asshole.

After the video stupidity, he says, "There are better relationships to be had."

Which he soon gets to experience firsthand. Kim is the CEO and founder of TravelingMom.com, a travel website for families. As part of her job, she gets to go on plenty of junkets, and whenever possible, she brings along her youngest child. During the first weekend of December, she takes him on a short cruise out of Miami, where he's seated at a table with some elderly Italian folks. Romano seizes on the opportunity to talk to them in his burgeoning Italian. He has a special fondness for older Italians because they remind him of Nanni. This, he says, is what matters. Not viral videos about a busted high school romance.

But there is something else that proves to be a salve to his wounds over the ongoing breakup fiasco. Part of the draw of the cruise is that Kim and Romano can stop by the University of Miami and pay a visit to Dario, the family's firstborn.

Dario and Romano have shared a bedroom their entire lives and there's definitely a Felix/Oscar dynamic at play, Romano being the fastidious one. Like Sophia, their sister, Dario's eager to dish on his younger brother.

"Have you heard about how he likes to scare the crap out of everyone in the family?" says Dario. Romano has an obsessive habit of trying to spook his family. It usually involves the most basic of gags, nothing more advanced than something you'd see on *Scooby-Doo*—jumping out from behind corners, lurking in the pantry until someone opens the door to find him standing there, silently creeping up behind someone until they turn around to find him nose to nose.

"Ugh, it's so annoying," says his father, Rome.

"ANNOYING," says Sophia.

"It's super annoying," says Kim, the only one in the family that Romano can't spook. She just looks at him, stone-faced, shakes her head, and walks away. Kim's a tough chick from Kentucky. It would take a lot to startle her.

The added feature of the Romano scare is that after you scream or jump, he immediately barks, "One to ten? One to ten? Come on, one to ten?" insisting that his family rate the fright factor.

"I have to say, there was one time when he totally got a ten," his brother concedes. Dario got up in the middle of the night and while he was gone, Romano crawled into his brother's bed and pulled up the covers. When Dario, bleary-eyed, lifted them, Romano reached up in the darkness and grabbed his arm. Dario screamed and jumped back.

"It was a whole other level," admits Dario. "I almost peed myself."

Despite the mild form of terrorism he unleashes on his family, his siblings are deeply fond of Romano, but they in no way coddle him. In fact, picking on him is one of their favorite sports. There's certainly no shortage of material. They also freely admit he's the academic ace of the family. Sophia says she "fully plans to live off" her younger brother one day.

When mother and son turn up in Miami, Romano hits a teenage jackpot that is the stuff of dreams. It happens to be the weekend of a sorority formal. And somehow there is a sorority girl in need of a date. And somehow Dario volunteers his little brother. And somehow the sorority girl is cool with this. Which is how it comes to be that Romano winds up at the formal, a junior in high school, surrounded by a bevy of college babes.

"It was incredible," says Dario. "I saw the photos afterward, and Romano is in, like, every single one. It's like, there's Romano, look, there he is again. Unbelievable."

He's got his mojo back, and when he comes back to the lab after the Miami jaunt, you practically need sunglasses to shield yourself from the glow emanating from Romano.

Andy's been increasingly irritated that he hasn't been able to repair the FTIR, the instrument that produces a chemical fingerprint of substances. So one night after school, he heads to Sacred Heart and runs Romano's two samples—plain gelatin and gelatin plus L-DOPA—through the university's FTIR.

He brings the printouts back to the lab and he and Romano sit down and examine the graphs. Each page has a single swooping line, like something you'd see on a polygraph test.

Right away, Romano looks suspicious.

Andy points to the remarkable similarity of the two graphs. "If you look at these two fingerprints—"

Romano finishes his sentence. "They're basically the same. So, the L-DOPA isn't making any difference?"

"Well, not from a chemical standpoint," says Andy.

After consulting with Andy and Bruce Lee, Romano decides to design a series of head-to-head tests between plain gelatin and gelatin plus L-DOPA—all with the goal of trying to unlock L-DOPA's purported adhesive properties.

The first is a water degradation test, meant to see if gelatin alone breaks down faster than gelatin with L-DOPA. If so, this might be because the L-DOPA is making the gelatin more stable—which would bode well for his liquid bandage.

Meanwhile Andy's been rotating parts in and out of the FTIR and consulting with his former colleagues. In an incredible stroke of good luck, he gets a call in late November from a buddy of his who has a brand-new FTIR, one that was taken out of the box and can't be resold. He gives it to Andy for free. Within days, there's a new $40,000 instrument in the lab.

Romano is one of the first to use it. But when he runs his water samples through the FTIR and prints out the chemical footprints, once again, the graphs are nearly identical. In other words, L-DOPA doesn't appear to be doing a damn thing.

His next experiment is much cruder and involves basic gravity. Romano places the boogery blobs on glass slides and does a drip test where he holds them vertically to see which one is stickier.

"None of them slid off," he says.

Verna and Romano lately have turned up in the lab at the same time to work on their projects. They chat and joke, Verna definitely being the straight man to Romano's boisterous display of Romano.

After seeing the drip test, Verna says to him, "Didn't you say it needed to adhere to a WET surface?"

True, says Romano. The gelatinous masses need to be able to stick to wet surfaces—to mimic wounds. He combines gelatin with water.

"No stickiness," he says. "In a wet environment, it's useless."

In early December, Romano's in the lab and microbiologist Prem Subramaniam shows up to work with his son, Rahul, on Rahul's Zika project. Prem is happy to chat and consult with kids who might benefit from his knowledge as a research systems biologist.

Romano had briefed Prem on his project during an earlier visit. But now that he's running into all these obstacles, he's eager for Prem's advice.

Prem tells Romano the combination of L-DOPA and gelatin likely isn't enough to do anything. He suggests that Romano add iron chloride to help the concoction solidify more rapidly. Iron chloride, he explains, is an oxidizer, and oxidation is ultimately what will affect curing time.

This makes perfect sense to Romano. He is, after all, killing it in AP Chemistry this year, with an average in the high nineties. And he's open to pursuing just about anything at this point. So in short order, he's adding iron chloride to gelatin alone and gelatin plus L-DOPA.

Romano is chronicling his experiments with photos and detailed notes—which he'll use for his midterm presentation and to have at the ready at any competitions. Here's what he writes about the iron chloride experiment: "This experiment did not end up supporting my hypothesis. In both trials, the solutions that contained only $FeCl_3$ (iron chloride) cured faster than the solutions that contained $FeCl_3$ (iron chloride) and L-DOPA."

Alas, foiled again.

"So Prem's idea didn't work either," he says. He admits that part of him feels vindicated by this latest fail. Hell, if an Ivy League researcher can't get L-DOPA to do its thing, how is Romano Orlando supposed to work any magic?

He gets back in touch with Bruce Lee. No, no, no, says Lee. You're adding iron chloride before you know what L-DOPA is doing.

"He said this is all very qualitative and complex, that I need to figure out L-DOPA first. I was like, 'Okay, yes sir,'" says Romano.

Even though he's sock sliding all over the place, skidding from his own instincts to those of the competing scientists, Romano doesn't seem deterred. In fact, he's still gleeful every time he says "L-DOPA."

Every year, Andy organizes a holiday gathering for the research crew. In years past, some research parents have opened their homes for the party. (One host family had a home movie theater.) No takers this year, which Andy finds to be a bummer. Now he's got to cobble something together. He calls around getting prices for his head count and decides the party will be dinner at Dinosaur Bar-B-Que, a place in Stamford. The kids will have to pay twenty dollars apiece.

On December 16, the gang trickles in and take seats at a series of tables that have been moved together to form one long massive table. Nearly all of the forty-eight kids make an appearance. The mood is festive and the kids are surprisingly jolly—something about getting out of the confines of school elicits a kind of energetic surge.

Romano's sitting across from Collin Marino, one of his lab table-mates. Collin breaks out a massive magnet and rubs it against some utensils, which somehow transfers the magnetic connection in such a way that he can make a fork and spoon dangle from each other.

Romano's kind of mesmerized and mystified. "Wow, you just had the magnet in your pocket?" he says. Who does that? He takes a few photos and decides they merit a little social media. He posts the images on Snapchat, along with some others from the dinner.

After school on Monday, Romano's in the glass corridor, the main artery between the student center and the music and science wings.

The corridor is nearly empty. He bumps into a Korean-American kid who apparently saw Romano's Snapchat photos from the barbecue joint.

"Why are you hanging out with those losers?" he asks Romano.

"Fuck you," says Romano. "You don't know those kids. I'd rather hang out with them than you any day."

He keeps moving, not wanting to engage in this stupid conversation.

What he doesn't know or see is that there's a witness to this exchange: Madeleine Zhou, the often caustic, sassy ringleader among some of the Asian-American research kids. No one has a more lethal combination of angelic face and tongue dipped in moonshine.

Madeleine quickly tells William and others what she saw—that Romano, unaware of her presence, immediately defended their people, the research kids, the ones considered by some to be losers, nerds, untouchables. It would have been easier for Romano to say "Whatever" or even join the heckler for a few seconds just because that's what people in the high school gulag do.

But no. Romano made his loyalty clear. It would be hard to overstate what this does for him in terms of his social standing with Andy's kids. Romano doesn't know Madeleine is spreading the word. In fact, the whole incident prompts Madeleine to declare, in a rare moment of zero irony, that "Romano is one of the nicest people." William starts recounting the incident, with no shortage of amazement and near disbelief at how Romano stepped up.

Many of the kids always knew Romano was cool. He's undeniably in the cool-kid group and does cool-kid things, like go to parties and get voted homecoming prince. But this moment in the corridor is where all the artificial social constructs bust open, when there is truly nothing to be gained by doing the right thing, and in fact doing so only opens him up to being ostracized.

In that unexpected, spontaneous moment, he's shown how cool he really is.

WINTER

29

Andy

Andy opens the lab a couple of days over the holiday break—both for the kids and for himself. This is his physical and psychological nest, the place that, even when beset with chaos, is a source of comfort.

On January 2, he calls four of his powerhouse girls—Olivia, Sophia, Dante Grace Minichetti, and Devyn Zaminski—and invites them to come in and work. It's a salve for all of them. These are the research veterans who can toil independently and who appreciate having the lab and Andy nearly to themselves. It's looking like they're all headed to the CT STEM Fair. They've gotten ahead of schedule, have been working hard, and Andy wants to reward them with a chill day in the lab where they can spread out, relax, and work without the distractions of a roomful of kids.

He puts on his music—soft, ambient electronic tunes that are easy to work and think to. In a perfect world, he'd be able to do more days like this one where he can tend to a small group of dedicated kids, getting to the meaty parts of their projects without being yanked away every few minutes by someone else.

Olivia brings him a sandwich, which always gives Andy a comical amount of pleasure. The kids know that to feed him is to earn goodwill. He doesn't expect or assume it, but he loves it when a kid recognizes that he's sacrificing precious days off and hands him a hot meatball hero on a cold day.

"It was a great day," Andy says afterward. "It was relaxed and fun."

Which he needs because when school resumes on January 3, he's about to enter into the vortex. The science fair season is officially looming and there will be no more quiet days in the lab with just him and his stalwarts. No, this is where the tornadoes start to swirl around his feet and work their way up.

On the first day back, Andy stands at the front of the room and, following some pleasantries about the holidays, tries to deliver a sobering message. Many kids have spent their breaks in Europe and the Caribbean and look very well rested, even tan.

"Keep in mind, this is a whole new routine. Generally speaking, I'm losing my flipping mind to get you to the state fair. So I don't know how this is going to happen. To be honest, there's still too many of you. I did a little perusal of who's working and there's still twenty-two of you. Only eighteen of you are going.

"People that have been here on weekends get priority. So if that's not you, get your s-h-i-t together. I haven't done the red, yellow, green thing, but that tends to light a match.

"There's so little time. I know midterms are coming. Depending on your test schedule, there's lot of [free] time in there, if you can fit it in.

"For the CT STEM folks, that takes you into the fourth week of January. For the other people, you have basically a month to finish up.

"Finish up means a week before. I need to get posters to those people a week to ten days before the fair. I start emailing them to the printer one or two at a time, but there will be a drop-dead date.

"That's it. Speech over. It's very cleansing."

He takes a deep breath and a rare silence falls over the room.

Senior Adam Roitman says, "Regeneron is tomorrow." Meaning the semifinalists will be announced.

Andy waves this off and tries to downplay the significance of the day. Eleven years in, he knows not to shoot off air guns or display outright excitement—because for any winners he may get, there will be those left behind. It's one of the hardest aspects to his job, simultaneously rejoicing for those who win and comforting the ones who don't.

Andy grows hot to the touch when parents complain to him that their kids lost, or, worse, insinuate that he doesn't get it. He gets it all right.

Sofia Bramante's public high school in Fairfield didn't have a science research program. But being raised by two scientist parents, she defaulted to science as an interest (she's majoring in mechanical engineering), and Andy wasn't about to let Sof be a case of the shoemaker's daughter. So he mentored her himself. Sofia engineered a color-changing synthetic skin, a unique form of camouflage that could have applications in the military. The project arose out of a conversation she had with a helicopter engineer. Andy was both proud of and wowed by his daughter's project. Like Andy, Sofia is a hard worker, she puts in the time and doesn't lollygag. Like Tommasina, she means business.

As a sophomore, she had developed an air purifying system to get rid of secondhand smoke. It was a good starter project but fell short of earning her a spot at Intel.

With her synthetic camouflage skin, she'd kicked it up several notches, and Andy thought Sofia stood to go far. Sofia badly wanted to get to Intel. How could she not? Since she was eight years old, she'd lived in the starry glow of her father's crop of winning kids, glitzy trips, and prize money. Naturally, she wanted to be part of it all.

So at 2013 CSEF, the two were dumbfounded as they watched, one by one, the top awards go to Andy's research kids—and Sofia wasn't among them. Their shock turned to heartbreak while Andy posed for pictures with his troupe of Intel-bound kids, trying to smile while he could see his daughter standing on the side with tears in her eyes.

Andy was tormented by the loss. Had he somehow dropped the

ball with his own daughter? How could all of these other kids tri-umph while Sof didn't make the cut? Even with all of the subjectivity and unknowns on the competitive circuit, this seemed especially cruel. He couldn't help but blame himself.

When they got home, Sofia sat on the couch and wept. Andy's cell-phone rang. It was a mother who launched into a bitter tirade about how he had not properly vetted the projects because had he done so, he would have seen that her child's project was too similar to another research kid's, thereby squashing her child's chances for a win. He had no capacity to take on the crazy mother. On a good day, he was aghast at this kind of entitled behavior. But on a day where he and his kid were hurting so badly, he could offer nothing to this woman. He quickly and pointedly told her he had his own problems and hung up.

That year, while Andy was at Intel with his winning students, he had a pit in his stomach and tears in his eyes as he walked the giant exhibition hall and could think of nothing and no one but Sofia. He knew he was supposed to be basking in Greenwich's unbelievable showing, but he could not rid himself of the feeling that he had failed the kid who mattered to him most: his own.

So while the seniors are twitching with anticipation to see if anyone is named a Regeneron Science Talent Search (STS) semifinalist, or "scholar," Andy remains staid.

Regeneron is the most prestigious science research competition for high school seniors in America, the Nobel Prizes for young people. It's also where the most money is to be won and has been known to propel winning kids forward with college admissions. Started in 1942, STS was sponsored by Westinghouse until 1998, when Intel became the title backer. Intel dropped its sponsorship in 2016 and now biopharmaceutical company Regeneron, with a market cap of $35 billion, is taking over. The company has pioneered treatments for everything from high cholesterol to macular degeneration. Its spon-sorship is especially meaningful because two of its top executives—

Len Schleifer and George Yancopoulos, both M.D.-Ph.D.s—were STS winners. Len was a semifinalist in 1970 and George was a finalist and top winner in 1976.

Applicants complete an extensive application, similar to a college admissions form, that includes their research project, transcripts, recommendations, essays, etc. From the pool of about 1,800 seniors, 300 semifinalists are named, and from that, 40 finalists are chosen. The semifinalists win $2,000 with an additional $2,000 going to their school. Which in Andy's case means money toward his research fund.

The 40 finalists are invited to Washington, D.C., for five days of competition, which involves Mensa-like problem solving and intense interviews. The finalists win $25,000 but then vie for larger purses awarded to the top ten winners. Those figures range from $40,000 for tenth place to $250,000 for first. (The top ten winners receive their earnings in lieu of the $25,000.) So a top win can essentially cover the cost of college.

The semifinalist announcement is slated for nine A.M. Several of the kids have their laptops open.

"I'm not going to check," says Andy, who's now walking and talking so fast, it's obvious he can't quite completely detach from the moment.

At nine A.M., the kids check the site and . . . nothing.

Seven minutes later, still nothing, and Andy underscores his position.

"I'm not going to check," he says.

But now the results are up. The kids gather around a few screens and read the list, which is alphabetical. Someone else finds the online booklet, which lists the semifinalists by state.

Greenwich High School is the first school under Connecticut and there are . . . six names: Olivia Hallisey, Ethan Novek, Sanju Sathish, Derek Woo, Devyn Zaminski, and Madeleine Zhou. Greenwich is the clear winner in the state. The next winningest school in Connecticut, Amity Regional High School, has three semifinalists.

This is Andy's second-best showing ever. During the 2014–15 year, he had seven semifinalists, all boys. But despite the gilded results, it was a tainted year for other reasons. There was a mean-spirited, sadistic current running through the class that year, with accusations of cheating that were countered by claims of sabotage and even an attempted sting operation. It poisoned the experience for many involved, Andy included.

By contrast, this year's seniors are a much tamer and kinder bunch, less selfish and savage in their pursuits, making for a more respectful environment. But Andy can't pause for much reflection at this very moment.

Senior Agustina Stefani checks her screen and looks grief-stricken.

"I just want to win SOMETHING!" she exclaims. "All these years and I haven't won a single thing. Ugh." She's been with Andy since her sophomore year. The project she submitted was a carbon capture device where she was able to engineer silver nanoparticles in water to trap carbon dioxide.

Andy goes over and hugs her, getting bunched up in her hoodie. Agustina's glum as she reads the names of the winners.

Andy hasn't looked at the list and didn't seem to register the names when the kids were saying them aloud. He stands there with Agustina and says, "No William?"

"I'm surprised not Yin," she says.

"Weird," Andy says and then lets slip, "He was the only one I expected to get it."

Someone says, "Danny?" which elicits a chuckle because by now everyone knows he didn't come close to making the deadline.

Andy says, "Margaret? I thought Margaret had a shot, too." He snaps out of his puzzlement and says, "I'm sorry, Miss Stefani. How's your other stuff working out?"

When the bell rings and the kids file out, Andy says, "I thought for sure W.Y. would be on that list. Olivia's still rocking the Ebola train. And Derek finally gets the recognition he deserves."

Besides Shobhi, Derek was the only other one of Andy's kids who wasn't named a finalist at last year's CSEF. His project was novel and different from anyone else's in the class. He'd heard a story on NPR about colony collapse disorder (CCD), the mysterious breakdown of bee colonies that has been leading to a reduction in the bee population, which in turn has implications for agriculture that relies on bees for pollination.

Derek determined that pesticides used for certain vegetation collect at high concentrations in the droplets that are emitted on the ends of the crop leaves, something known as guttation. When bees were exposed to the pesticides in the droplet, if they didn't die outright, it made them disoriented and unable to find their way back to the hive—hence the collapse of the colony.

Derek introduced charcoal produced from plant matter (known as biochar) to see if it would raise the threshold for toxicity that was disorienting the bees. It worked—the biochar brought down the concentration of the pesticides, which theoretically would make the bees less dizzy.

Andy was shocked when Derek didn't even make it to the finalist round at CSEF, especially for such a timely, cool project. So he's especially chuffed to see Derek's name on the Regeneron list.

Plus, poor Derek hit one of the weirder and more jarring tripwires in the whole college admissions field of land mines. A top student and competitive rower, Derek applied early to MIT, which does not require early admission applicants to attend if accepted, and an Ivy League school, which does. (In other words, if Derek were to be admitted to the Ivy League school, he would have been required to accept that offer.) Derek says that the MIT crew coach contacted him and strongly implied that he would be admitted early—on the condition that he withdraw his application to the Ivy League school, where he's a legacy. Derek agonized over the decision but eventually decided to pledge allegiance to MIT. He reluctantly withdrew his application and convinced himself it was the right thing to do. So when MIT's early notification day came and he was passed over, he and his family felt completely duped. Derek's mother, Ryeo-Jin, was distraught for days. Derek walked around in a haze, struggling for words to explain the situation to everyone who assumed he was in, given the overtures from the crew coach.

Andy's next research block starts trickling in. When Derek learns he's a Regeneron semifinalist, he smiles for what seems like the first time since the MIT debacle.

Andy is also really happy that Devyn Zaminski has been recognized. Andy always says that Devyn is the "whole package," because she's smart, hardworking, humble, and kind. She's also a heart-stopping half-Caucasian, half-Japanese beauty. Despite growing up in a comfortable family, she says her parents are always telling her that Greenwich in no way represents the world at large, or even a sliver of it. Other kids often mock their ritzy hometown, but it can feel canned, like it's the thing you're supposed to do to insulate yourself from looking like a clueless brat. But somehow when Devyn talks about this aspect of life in Greenwich, you know she gets it.

Some of this is from volunteering over the past several summers in Guatemala, which helped inspire her project. She created a self-contained energy system that uses a microbial fuel cell (made from poop) to clean water and in so doing excretes methane gas that can be used as an energy source. Her time in Guatemala made her want to do something sustainable that could be deployed in developing countries. When Devyn took her project out on the science fair circuit her junior year, the judges were awed. "When I was at the fairs, a couple of judges were asking me to patent it," she says. "I'm thinking about it."

Sanju walks into the lab and immediately slumps down on a stool. He has no knowledge of the news and is in a general funk. "I'm really tired. I hit the snooze button twice," he moans. "It just ruins the rest of my day." He's wearing sweatpants, looking very much like he's not fully aware that he's left the house.

Madeleine Zhou walks in, also unaware. She opens her laptop to the Regeneron page and immediately types in "Yin" to the search field. When nothing comes up, she starts frantically refreshing the page.

"Shit!" she exclaims. "Wow."

Sanju has since seen his name and says, "Cool." He's finally starting to wake up from his sleepy stupor and has moved on to the more shocking development. "Can you believe that about Yin?"

"It's okay, he has so much other stuff," says Madeleine. Still clueless that she's won, someone tells her to search on "Greenwich" instead of "Yin." After six taps on the Enter key, the highlighted text lands on her name and Madeleine nearly topples off her stool. Her project, titled "Selective Improvement of Anti-Cancer Drug Sensitivity via D-Glucosamine Inhibition of STAT3 Oncogenic Expression," delved into ways to increase efficacy of chemotherapy in cancer patients. Given that Madeleine's default state can be glass half empty, she's genuinely surprised.

Madeleine and William have a complicated dynamic. At times she can be caustic and fiery and brutal to him. She mocks and trash-talks him. Other times, she simmers down and will play online games with him and appear friendly. And still other times, it appears that despite herself, she is rooting for him. Overall she has a magnetic love/hate relationship with him, his academic reign, and everything he stands for.

For his part, William takes whatever he can get from her but clearly seeks her approval. When she goes cold, he retreats like a dog who's just been smacked on the nose with a rolled-up newspaper. And when she's warm, he basks and lolls about in a semi-euphoric state. But always, he wants to be in her good graces. Of course, any trace of apparent weakness on his part seems to infuriate her and set up opportunities for exploitation. It's not unlike the scenario between Lucy and Charlie Brown where Lucy's always setting up Charlie Brown to kick the football and then pulling it out from under him at the last second. Seeing it play out in person, William's eyes going all doelike when she's feeling benevolent toward him, only to position himself for the inevitable smackdown, is both comical and head-scratching.

Everyone from Andy to his father has told William to quit being so submissive with Madeleine. It's beneath him and a waste of time.

But today, she appears to be the slightest bit disappointed on William's behalf, because it just doesn't seem right that he was omitted. William has stayed home "sick" today, which means he's working on his college essays. He's also being spared the public spectacle of his Regeneron slight.

Andy is undoubtedly happy for his six semifinalists, and of course he's pleased with the fact that his research coffers will now have an additional $12,000 for next year, significantly decreasing his stress around the money.

But when everyone's gone and all the congrats and commiseration have been doled out, he lingers on William. He doesn't go through the mental torture of trying to think through all the plausible explanations. He's too seasoned for that. Instead, he just thinks about the salt on the wound aspect.

"The Harvard thing and now this," Andy says, sighing, shoulders down. In these moments, the creases across his forehead seem deeper, as if the vagaries of life are digging in and taking hold, limning a map of stress. "This would have been a good thing to lift his spirits. What can you do?"

Regeneron allows the three hundred semifinalists to luxuriate in their glory for a few weeks before they announce the forty finalists. That's slated for January 24.

During the 2015 school year, when Andy had his personal record of seven semifinalists in this competition, not a single kid advanced to be named a finalist. Given all the animosity and vitriol among the kids, it seemed like karma intervened.

In fact, over the past ten years, only two of his kids have been named finalists, which shows just how insanely competitive Regeneron is. Stephen Le Breton made it in 2013, Annie Merrill in 2014. The Merrill family invited Andy and Tommasina as their guests to Washington, D.C., to attend the grand ceremonies. Annie got to shake hands with President Barack Obama. The whole experience was a career highlight for Andy—to watch one of his kids ascend to the highest echelon of the competitive science fair world.

But this year, for some intangible reason, Andy says, "This Regeneron thing . . . I'm not feeling it."

Andy and Tommasina are standing in their kitchen after work and Andy echoes his sentiment to his wife. "You know, I remember that with Annie, they called her the night before with the news. I'm not feeling it this year, Tom."

No sooner do the words leave his mouth than his cellphone rings. He looks down. Ethan Novek.

"Ethan! My friend!" says Andy.

Ethan practically leaps through Andy's phone with excitement. He's been named a finalist, which, in addition to the prestige, also comes with a $25,000 cash prize. Despite his long track record of winning competitions, Ethan is genuinely ecstatic.

He and Andy chitchat for a few minutes and when they hang up, Andy checks his messages. About twenty minutes earlier, Derek had Facebook-messaged him that he, too, was named a finalist.

Ok, *now* Andy's feeling it. Two finalists in the same year! Two fantastically deserving kids, including one who'd been overlooked the year before. Ethan and Derek are the only finalists from Connecticut.

"Oh, man," says Andy. "This is great news for them."

The next day, the lab is humming with activity. The local newspaper arrives at school to interview Derek and Ethan and take their pictures.

Ethan walks in at 2:05 P.M. and is showered with warm greetings. People want the blow-by-blow of how he got the news. In what's an increasingly rare occasion, he was actually at his family's home in Greenwich.

"I saw the 202 area code and thought it could be the patent office," says Ethan. (As you do.)

Derek was in his bedroom practicing his clarinet when he got the call. He says there were a bunch of people assembled on speaker-phone yelling, "Congratulations!" He was so shocked, he couldn't muster much beyond a thank-you.

The photographers start putting the boys in place for their photos.

"Smile!" says Andy with a laugh. "Look at those shit-eating grins."

The conversation then turns to the big finalist week in D.C.

"You guys get to shake hands with Donald Trump, by the way," Andy points out. The kids in the room start laughing and discussing whether such a thing is actually an honor. Some question whether Trump would even show up at something that seems so far afield from his interests.

"If he doesn't," ventures Ethan, "that would be such bad PR."

"I'd still want to meet him," says Derek after thinking it over for a second.

The consensus seems to be that meeting any sitting president is rare enough so as to be cool, but it would mean more were it not the buffoon who is the current leader of the free world.

The day after the Regeneron announcement, Derek pulls Andy aside to chat. Madeleine told Derek that someone made disparaging remarks about him on a Chinese news site that ran an article about the forty finalists. The anonymous comments beneath the article are detailed and appear to be written by an embittered parent closely tracking the results of Andy's kids.

Andy is none too pleased to hear this. He has grown somewhat de-sensitized to Greenwich parent behavior, though just when he thinks he's seen it all, some parent will outdo the pack. In this case, he doesn't want to jump to conclusions. Derek hasn't seen the comments—and, being Korean and unable to read Chinese, can't understand them any-way. Andy says he'll get Madeleine's version of events. He is taking it seriously but trying not to inflame the situation.

Privately he says he wouldn't put it past any parent in Greenwich or Fairfield County (the poshest in the state) to do something like this.

Derek presses Madeleine for more detail and learns that she never actually saw the commentary. It was relayed to her by her mom. When Madeleine looked for it, the comment had been deleted after someone responded to it.

Here's a translation of what the comments said:

> This year's Regeneron STS Finalists don't make any sense. Two kids in our school are in the finalists, one of them won first prize in Intel, got great grades in school, and got an offer from Yale. The other Chinese boy only won a local prize and his grades are only above average, he didn't even win a prize at the Connecticut State Fair.
>
> There is another Chinese boy, his GPA is the highest in school, he won second prize in Intel for two years, and other prizes, too, and he's not even a semifinalist. Another American girl won first place in Intel and other famous science competitions. She's not in the finals, either.

The comment was written by "May Mom."

To reiterate, Derek is NOT Chinese and his grades aren't "only above average."

Derek is a complicated blend of hurt, angry, and intrigued. Why would an adult score-keep to this degree and then be so petty as to publicly disparage someone else's child? Even if it is on a Chinese news site that few people he knows can or would read? To Derek and Madeleine, it seems clear that the culprit is a Chinese Greenwich mom who closely watches the performance of the science research kids. This narrows the pool significantly, but Derek has no interest in trying to actually prosecute the crime. He points out that the absence of hard proof means any suspect can simply deny it—but once the accusation is out there, things will be extremely awkward.

Still, he doesn't immediately move on.

"I was shocked. It wasn't a good feeling," says Derek, who has a look of disbelief as he talks about it. "I don't see how it would achieve anything."

30

William

What could be worse than not getting into Harvard, the one place his family had hoped and dreamed for? For William, not being named a semifinalist for Regeneron STS is a close second.

It's a sucker punch that lands with a particularly knuckling blow because advancing in Regeneron is so prestigious and elite, it often adds the extra sparkle that pushes someone into the college acceptance pool. Andy has seen this happen a few times with former students.

It all just redoubles William's focus on his new project, the Alzheimer's diagnostic, for which he's now created a broad template.

Andy stands back and waits for William to come to him to discuss things before offering his thoughts. For months, the most specific William could get is that he wanted to do something brain based. So when he finally walked into the lab in early December with this idea, it felt like a moment.

He plunked down his backpack and said to Andy, "The area I'm moving toward is diagnosis. I'm interested in a paper-based diagnostic." He paused and then added, "I've got the feeling that this project is really out there."

"That's what you said last year," replied Andy.

"It's like the feeling of last year when I thought I'd use the magnets," said William. He was referring to his atherosclerosis project where he tinkered with the idea of people ingesting a pill containing iron nanoparticles that would bind to arterial plaque. A handheld nuclear magnetic resonance device would be held to the neck to show whether a magnetic field was present, signaling plaque. William went so far as to produce a series of diagrams detailing his plan, which was admirable, but cuckoo in terms of viability. It could live on a shelf next to Leonardo da Vinci's flying machine. Even William now concedes, "It was really out there and not very practical."

Andy listens to William as he lays out the basic idea for his Alzheimer's diagnostic: You'd place two classes of liquid chemical compounds (what William calls "Class A" and "Class B") engineered into nanoparticle vessels (think super-tiny protective boats) under your tongue (sublingually). The liquid compounds would travel to your brain via the carotid artery.

The Class A boat contains gold nanoparticles and horseradish peroxidase, an agent that will produce a color change on a urine-based test strip if Alzheimer's is present. The outermost layer of the boat is curcumin, a naturally occurring plant substance that has been shown to cross the blood-brain barrier. (The gold nanoparticles don't serve a diagnostic function, they're merely anchors.)

The Class B boat contains gold nanoparticles covered by curcumin and tetrazine.

Imagine these two invisible lifeboats coursing through your bloodstream. Beta-amyloid is one of the signature protein fragments found in the brain when Alzheimer's is present. Beta-amyloid forms molecular complexes called oligomers, which are telltale signs of the disease.

The whole test hinges on whether the lifeboats encounter an oligomer. If they do, the boats will moor themselves to the oligomer, set-ting off a series of reactions. The tetrazine in the Class B boat destabilizes the environment and breaks apart the Class A boat. Horseradish peroxidase is released into the body, producing a color change on a urine test.

"The reaction is binary," says William. The only way this chain of events happens is if an oligomer is present. If it's not, everything re-mains as is and the compounds would be excreted from the body within forty-eight hours.

William has whipped up this little idea 100 percent on his own. He's not working with an Alzheimer's researcher. He's never done an internship involving the neurochemistry of Alzheimer's. What he says he has a knack for is the ability to read the literature and see all the disparate aspects of the collective research and then stitch them together to create a novel process.

"It's not bad; the only thing is you lose control of the back end," says Andy, referring to the fact that the interactions between Class A and Class B feel a bit tenuous.

For his part, William would love to be able to have a single boat of compounds that reacts if it reaches an oligomer, instead of two boats and a series of reactions. He's mulling over whether he can engineer this into a single nanovesicle.

"There's a lot of uncertainty," William concedes.

"You can't pull this stuff back through the skin?" asks Andy, won-dering if there's some other way to obtain the final result.

"None of the particles go in the bloodstream, like with atheroscle-rosis. The bloodstream is so turbulent, so it has to be saliva or urine."

"Tears, maybe?" asks Andy. "Is there anything you can get your hands on? It's not a long way to go from blood to your eyeball."

"Amyloid builds up in poopies," offers William, which prompts a face scrunch from Andy—though Andy must be pleased to hear William deploy Andy's term of choice for fecal matter.

They go back and forth, discussing the limitations and virtues of William's idea.

William points out another benefit: its low cost. He estimates the test could be manufactured on a large scale for a dollar each.

"If this were to be developed for developing countries . . . ," says William, already thinking big.

At the end of their chat, Andy says, "You're committed to this topic, which is commendable."

"It's interesting but so difficult with so much uncertainty," says William. "The biggest limitation is I can't do anything of this in vivo." In vivo being in living organisms. From federal regulations to science fair ones, there's no way that a seventeen-year-old can run tests in animals or humans. So this means that William is left to perform a series of proof of concept experiments that with luck will show that each aspect of the project is viable.

On January 27, William receives an email from the Junior Science and Humanities Symposium (JSHS), a STEM competition sponsored by the U.S. Army, Navy, and Air Force, informing him that he's been chosen to deliver an oral presentation at the event on March 11. Rahul and Luca Barcelo are selected to present posters of their work, which is still prestigious but a rung below orals.

William forgets to tell Andy. Shobhi and Connor Li are also chosen to deliver oral presentations. When Shobhi hears William's going, she says, "Oh, great, now the prize I might have won is going to Yin." She's incredulous at how blasé William is about advancing.

When he says, "Oh, I really forgot," Shobhi mockingly retorts, "*Oh, I'm so sick of achievement.*"

On a personal note, William seems to be branching out a bit—for better or worse. One of his regrets about high school is that he feels he should have been more socially proactive. He's in no way a loner; quite the opposite. In the smart kid sphere, he's popular, aside from disdain from some other Asian families in town.

But he wishes he were more socially forthright. He has the instincts to accept invites to hang out, but something, and he can't quite pinpoint it, often keeps him from acting. Speaking of which, he's decided, of all things, to audition for the spring musical, which this year is *Annie*. (William gets cast as the butler, which requires singing and dancing.) So he's taking some steps to remedy his less social side, as he did last year in a tortured bid to attend the junior prom.

William had missed the deadline to apply to ISWEEEP, a newer, global fair in Houston, but Andy made a plea with the fair, asking them to please take a look at William's project. They did and accepted him to compete. The awards ceremony fell on Saturday, May 7, 2016, the same night as Greenwich's junior prom. Andy had five kids going to ISWEEEP, most of whom wanted to leave early to make the prom. One girl told him she needed to fly home a full day before because air travel would make her swell up and she could not attend the prom in a bloated state.

Having secured William's spot, Andy made it clear to him that he needed to stay through the awards. For one thing, he stood a good chance of winning big. For another, Andy knew it looked snooty and disrespectful for the kids from tony Greenwich to jet out when it suited them, not even sticking around for trophies and photos. So Andy drew his line in the sand and William was game. He'd stay for his presumptive victory lap, easing Andy's anxiety. There is a degree of politicking in the science fair world, and Andy strives to keep Greenwich's reputation clean.

But two weeks before he was due to leave for Houston, William messaged Andy, saying he needed to talk to him. They agreed to speak that night. Sometime after nine P.M., Andy's cellphone rang.

"Hi, Mr. B."

"Hi, William. What's up?"

"Well, I was wondering about this issue of leaving ISWEEEP early, if it would really be a big problem."

"What do you mean?"

"Well, like, if I'm not at the awards ceremony, is it going to be a big deal? And if I knew I wasn't going to be there, would I have to tell the ISWEEEP people ahead of time?"

Andy could sense what was coming.

William said, "There's this girl . . ."

Her name was Emily, a senior.

"Well, for starters," said Andy, "do you know if she'll say yes? Because there's no point in us going through all these machinations if you don't even know if she'll go with you." Andy had a flashback to an earlier intervention when he stopped William from asking a girl to the prom whom he'd heard didn't want to go with him.

"I'm pretty confident she'll go," William replied.

"Okay, well, you have to do what you want to do. If you want to go to the prom and you think she'll say yes, well, then you should go." It was hard for Andy not to register his disappointment, but what was he going to do? Having cleared the path for William, he was ready to hang up and go to bed. But instead of William eagerly exiting the conversation, pleased with his furlough, he lingered. Andy reassured him they'd work it out. Still William hung on, uncomfortably so.

"William, use your words. Come on. Just tell me what you've got to tell me. It's almost ten o'clock. I can't stay on the phone all night."

William waffled on, saying he'd written out a list of pros and cons about whether to skip the awards and go to the prom. Eventually, he got to the heart of the dilemma.

"It's my image," he blurted out.

"Excuse me?" said Andy, trying to stifle his laughter. "And what image would that be? What are you talking about?"

"Well, I really feel I need to break free of this image of this nerd that's always studying."

"All right, let's play devil's advocate. Is going to the prom going to shatter that image for you?" Andy chided with affection and mild ball busting. "Is that really going to do this for you?"

"Well, maybe not," said William. "But for me it will. I need to do this. I need to put something ahead of what I usually put at the top of the list."

And this is when Andy knew that nothing that had happened in the lab meant as much as this conversation.

— — — —

After the December holidays, there's a rather noticeable wardrobe addition to William's otherwise predictable daily attire. Normally he wears jeans and sneakers and either a T-shirt or hoodie. No shirts with buttons or collars unless he has a concert. Like Steve Jobs, William's uniform means he doesn't ever have to lose a minute to thinking about what he wears.

So he causes no small amount of shock when school resumes in early January and he walks into the lab wearing a black leather (pleather?) jacket over his T-shirt. Think Members Only jacket from the 1980s, but in shiny black leatherlike material. For some unknown reason, he pushes the sleeves up, adding to the retro feel. He takes his usual spot on a stool at a high-top table, very much looking like someone trying to appear comfortable and normal, like there's absolutely nothing out of the ordinary happening.

It doesn't work. Pretty much no one can have a conversation with William and not acknowledge The Jacket. Some manage not to say anything directly, but the minute he's out of earshot, they ask, "Did you see Yin's jacket? What's up with that?" It's not even mean or mocking; it's pure bafflement, as if he'd walked in wearing lederhosen and an alpine hat. Some people go the philosophical route: "What do you think it means?" Others do that thing where a false, high-pitched compliment tumbles out of their mouths just to say something, anything: "Wow! I like your jacket!" Still others take a stab at its origins: "Christmas gift?" (He bought it on Amazon.)

A couple of days later, William appears wearing a second new black jacket, a Saks Fifth Avenue thin down coat, one of those quilted deals with the horizontal stitching. Unlike the leather jacket, the quilted down coat is much more befitting William. It's still noticeable simply because it breaks his standard uniform, but it's a smart upgrade. He looks sharp, a more fashionable version of himself, whereas the leather jacket is just distractingly out of character.

"Another new jacket?" kids ask. Some blatantly tell him it's a good choice and he needs to bury the leather one. "Much better," they say.

William seems to take in all the critique and The Jacket disappears after about a week. The quilted down coat remains.

On January 16, Martin Luther King Jr. Day, school's closed, but Andy opens the lab. Six early bird kids turn up in the morning. As the day plods on, a steady stream continues to walk through the door. Rahul's here working on the chemical aspects of his Zika project. Specifically, he's trying to determine the range of detection for apyrase, an enzyme in mosquito spit. (He's constructing the actual mosquito trap at home using materials from Home Depot.) Olivia's here working on her Lyme diagnostic—and is pleased to see it works without any hiccups. Verna's here, hitting somewhat of a wall with her project, a saliva-based test for Parkinson's disease. She's working with a chemical concoction that should change color if it's working. So far, no color change.

Just before two o'clock, William strolls in.

"The man, the myth, the legend," says Rahul.

Andy's making his rounds from kid to kid, checking in and offering assistance to those whose projects are sputtering.

When he spots William, he says, "William, I'm afraid to even ask about you. What's up?"

The two start discussing the mechanics of the Alzheimer's project. William isn't yet sold on his own idea. He's listing in trademark William Yin fashion.

"I have a concept that's very, very weak," he says. "I know that if I were to develop this further, I could make it stronger. Right now, the idea is not the best."

He's still unhappy with having two classes of compounds and a series of reactions. He's been reading a ton of papers and exploring every possible work-around, but nothing seems viable.

Andy hears him out, but his perspective is colored by so many whirls on the William Yin carousel. He knows William is a gold medalist in overthinking, which delays firm decisions, stalls action, and then pushes William so far up against the wall that he explodes into mania, pulling all-nighters, staying in the lab at the expense of everything else in his life. It stresses Andy out and irks him because he wants William to demonstrate that he prioritizes his research by conducting it in a calm, sane manner.

For his part, William feels he's getting a jump start on his Alzheimer's project compared to prior years. But it's all relative.

The kids who are here today are pushing to get their projects done for the CT STEM Fair, which is February 4, three weeks away. CSEF, the legacy state fair, happens in mid-March.

"I think I could bring the concept and some preliminary results to STEM," William says.

In theory, if he worked steadily over the next few weeks, it's possible he could produce some results, but at this stage, he doesn't even have any of the materials he needs.

Four days later, William takes the first step. He's on the phone ordering a thousand dollars' worth of chemicals. Andy's none too pleased because this order is essentially wiping out the research coffers.

"Second day shipping is one hundred seventeen dollars," William says, opting for standard, which means the chemicals will arrive in four days.

The next few weeks are frustrating. William is trying to simultaneously work on Class A and Class B, but he's still awaiting a compound he ordered.

The outer layer of both lifeboats is actually a polymer, molecular chains that when grouped together can form plastics and resins. The first thing he needs to do is chemically create this polymer. He reads a paper by a researcher at the University of Edinburgh and is completely spooked. The paper details a process so complex, William says simply, "Uh, there's no way. This is an organic chemistry experiment that's at such a high level and requires an isolated lab that we don't have. This is something even professional organic chemists wouldn't easily be able to do."

Fellow researcher Alex Araki overhears William and Andy discussing the project and gives William another paper about how to build a polymeric shell around nanoparticles, one he'd come across while casting around for ideas.

William reads the paper and gets inspired. He reads more papers on the chemical construction of a polymeric shell and settles on a version that isn't exactly the one outlined in the original paper from the University of Edinburgh but could still work.

One day after school, he's mixing his chemicals. He doesn't require much in the way of oversight, and Andy checks in but doesn't hover. After a half hour of mixing, he leaves his clear liquid in a beaker and heads home.

The next day, when he walks into the lab, he's met with a surprise—a somewhat horrifying one. A creepy gelatinous gray blob the size of a large walnut shell has formed in the beaker. It looks like a gag from a Halloween store.

"Um, yeah," William says when people peer at it quizzically. "It's, uh, not supposed to be a blob." You could make a funny video montage of people's faces as they look at the gooey mass.

It was supposed to have remained in a liquid state. William goes into triage mode, desperately trying to figure out how to salvage his thousand-dollar goo. He walks the beaker over to the fume hood and tries to dissolve the blob by adding a cornucopia of chemical solvents. The blob, fully executing its function as both joke and sci-fi prop, is impenetrable.

William tries poking it with stirrers. It's the consistency of a very firm pudding. It's not moving.

Over the next several days, he heads back to the literature and reads up on whether a mechanical intervention can break it down. He looks up possibilities such as centrifuging—high-velocity spinning. Will the centrifugal force split open the blob? He investigates sonication—

the use of sound waves to break it apart. Not surprisingly, he can't find any research showing that these methods can dissolve a weird blob formed in an attempt to create polymeric nanovesicles.

All the while, time's a-wastin' and he knows it. He later says, "Nothing worked. I could see the time dripping away. It was so disheartening. I thought it was an incredible idea and at the very first step, I thought I should give up. It was painful. I just kept thinking, how am I going to finish this project?"

The upshot: Yin fought the blob and the blob won.

But in the midst of his frustrations, a bright spot emerges. On Friday, January 27, as he's working after school in the lab, he gets a text from his father:

"You got into Stanford."

31

Romano

When Romano walks into room 932 following the holiday break, he's got a look of renewal.

"I spent a lot of time with the family," he says. "I made pizza one day with Nanni. I saw *Star Wars*, went to Costco." He also won his fantasy football league and pocketed $120.

For Romano, family time trumps all. He and some of his extended family went clay pigeon shooting in upstate New York, where he had some epic shots. Equally important, he'd curated his shooting outfit with great care—and felt he'd killed it.

"I had the England scarf, the Sorels, the corduroy. I threw the scarf over the shoulder. [Dario and I] were both really good, but I just looked better," he says, grinning. He whips out his phone to show a photo of him and Dario in the woods. He looks like a character out of a Wes Anderson movie.

Romano and Dario were competing against their cousins, with the losers having to buy the winners Chipotle bowls.

"I had one bullet left. I shot it into the air and just sniped it," says

Romano. "Oh, and it says 'Made in Italy' on the gun." This, for him, was an omen that he was destined to be a sharpshooter.

The Orlandos are part of one of the more interesting subcultures of Greenwich, the Rosetans. The Rosetans all hail from Rose, Italy, a small mountaintop town (population 4,365) in the country's southern Calabria region. Early in the twentieth century, Rosetans started immigrating to Greenwich and Stamford, primarily working as stonemasons and later filling other trade positions. There's no official data about the current population of Rosetans in Greenwich, but Romano's father, known as Rome, estimates they number a few thousand.

Rome's parents immigrated to the United States in 1960. He was born at Greenwich Hospital the following year. The family settled in the Chickahominy section of Greenwich, which had become a Rosetan ghetto. Many relatives followed suit. Rome suspects he has at least twenty second cousins just in town.

The Orlandos eventually bought a large home in downtown Greenwich where Nanni still lives and that serves as the family's culinary and holiday hub. Rome's younger brother, Joseph, and his Italian immigrant wife, Carla, live in an apartment connected to the house.

Rome grew up in Greenwich in the 1970s, as part of the true working class. His Greenwich was far from country clubs and hedge-lined estates.

"All my friends were Irish, Jewish, and black," he says. "I felt like an immigrant because there were so many first-generation Italians. You really didn't have the separation of wealth. There was always money in Greenwich, but those people drove a brand-new station wagon, they lived backcountry. I never realized the wealth of Greenwich.

"I didn't sense it till way later in the 1990s, the hedge funds, the Wall Street money. When I was growing up, the only thing open on the Ave. after six P.M. was a Baskin-Robbins. And then Woolworth's went out and Saks came in."

In high school, he ran with the bad boys. When asked if his posse were the kids who'd take down your mailbox with a baseball bat, he doesn't hesitate: "Definitely." Though he was in the tough crowd, Rome says he himself wasn't especially tough.

His father had always wanted his son to be an attorney, but what Rome really liked were cars and working with his hands. After he graduated from Greenwich High, he headed to Colorado to enroll at the Denver Automotive and Diesel College. He wanted to be a mechanic.

When he completed the program, even though he had no formal job experience, he decided to aim high. Résumé in hand, he walked into Greenwich's Ferrari dealership. The guy interviewing him saw his name and asked if he spoke Italian. "Yeah, I speak fluent Italian," he told him.

"There was an old Ferrari mechanic there who spoke no English. He hired me on the spot. It was the first time, I was like, wow, it really does help being Italian. We would travel all over the Northeast and I'm his right-hand man. We took road trips to work on Ferraris."

In 1981, when Rome was just twenty years old, his father died at the age of forty-one from a heart attack. His passing devastated the family. Even though Rome was quite content working at Ferrari, he was haunted by his father wanting him to go to college.

So he started taking night classes at Norwalk Technical College, where he got his associate's degree in engineering.

What transpired over the next two decades of his life shows the tenacity and work ethic ingrained in the Orlandos. Even bearing in mind he's of a different generation, Rome's trajectory stands in stark contrast to the college/achievement vise grip that's strangling many kids in America today.

While he was at Norwalk Technical College, an Italian manufacturer of computer numeric control equipment recruited him for a job out west, promising they'd later send him back to school to get his four-year degree.

Rome moved to California—his first time living anywhere outside of Connecticut. It was out there that in 1987 he met Kim, who was going to college in Tucson and was in California visiting a friend. They exchanged numbers and started a long-distance relationship on the phone. Three weeks later, Rome went to Tucson for a visit, and they've been together ever since.

He hadn't forgotten about finishing college. "I was done being a service technician," he says, adding, "I was always good with numbers." He took some accounting classes at Santa Monica College, a community college, and was later accepted to UCLA and USC. He chose USC because of its accounting program.

So in 1992, at age thirty-one, Rome enrolled in a four-year college for the first time in his life. He wasn't excited to go to class with a bunch of eighteen-year-olds—it was humbling. Kim had a different take. "You're going to be thirty-five years old with a degree or thirty-five years old without a degree," she told him.

He was right that he was good with numbers. He graduated magna cum laude from USC and was offered a scholarship to get a master's degree in taxation. By the time he was done and he and Kim were packing up their U-Haul for the move back east, he already had a job with PricewaterhouseCoopers in Manhattan.

"I knew I had to come back to Connecticut," he says. "I had a lot of family obligations. I wanted my kids to have some East Coast sensibility. I told Kim a long time ago, just so you know, I'm going back to Connecticut."

First and foremost, Rome wanted to be near Nanni, his mother. Despite becoming a widow in a foreign country in her late thirties, Nanni had decided to stay in Greenwich. Life would almost certainly have been easier for her back in Italy, but America offered her kids . . . America. Upward mobility, potential, dreams, all that. As for herself, with her limited English, she got a job working in the kitchen at Greenwich Hospital, where she had given birth to Rome. She sliced fruit and made salads there for thirty years.

In 1997, Rome passed his CPA exam. Today he's the CFO at a real estate development firm in town.

A first-generation kid turned Ferrari mechanic turned CFO—it's about as unorthodox a path as anyone in this town has taken. But this is why the Orlandos, despite being multigeneration Greenwich residents, in no way resemble any of the stock Greenwich characters. They have no airs, they open their home and invite people over for dinner without a second thought (this might seem basic, but it's notable how walled off and frosty people can be about their homes in this town), and they don't engage in any of the perverse status competition that constantly lurks here, the ever-present awareness of who has money and who doesn't. Their three-bedroom house, in central Greenwich, is on a quaint but unfancy street, and they converted their garage into an apartment for some rental income.

Put simply, they are among the most un-Greenwich people in Greenwich.

Kim finds all the pretenses to be both ridiculous and stifling and takes pride in her family's distance from the bourgeoisie.

"I don't fit in here," she says, no indignation in her voice. More like relief.

Romano revels in his Italian heritage. Hence his determination to learn Italian, to learn how to cure meats and jar sauce and play bocce ball with the *paesans,* the local Rosetans. He's aware that he lives in a town where some families urgently display their wealth, but he sees right through it, and if anything, it's taught him that money and affluence, like a bad oil leak, can have a poisonous spread that taints people. As a young child, Rome says Romano would ask him whether they could afford things before Rome would purchase them.

"I tell the kids, there are always going to be people who have more than us and there are always going to be people who have less than us," says Rome. "Don't judge your friends and people by how much money they have, judge them by how they treat you."

Sitting in the family's kitchen, he says, "You know, it's funny. Dario wants to buy the world, Sofia wants to entertain the world, and Romano just wants to change the world. He's definitely the most sensitive, the most caring, loving to both his siblings.

"All my kids love each other, but when you have conversations, he's the one who will say, I see your point. He can really take another person's point of view and see it. He's always been an old soul."

But his youngest son still annoys the living crap out of him with his fright antics (Rome concedes that Romano once got a seven at his expense) and some of his dandyish behaviors drive everyone crazy.

"Never go to CVS with Romano. He opens all the soaps and deodorants, has to smell everything. I once spent an hour and a half in Lord and Taylor with him while he's smelling all the colognes," Rome says, shaking his head at the memory. "He also likes to burn scented candles. And incense."

But in the scheme of things, Rome and Kim know they've raised particularly grounded kids in this status-conscious town. And Romano says loudly and proudly that no matter where he ends up, he very much wants to move back to Greenwich to raise his own family.

"I can't imagine being more than an hour and a half away from my family," he says. "I would just want my kids to have the experience I've had. I think this is a great place to raise kids."

Of course, any sweeping proclamations about life plans made at age seventeen are subject to change—often within the next ten minutes.

The same goes for matters of the heart. Romano's taken a laid-back approach to the ladies following the epic breakup with Caitlin. "I'm keeping my options open," he says.

There's an Italian-American girl, a senior, he calls "the most beautiful girl in the school." We'll call her Alessandra. He's so awed by her, he actually cuts a wide circle around her. He doesn't want to move in too aggressively and tip his hand. But he loves the fact that she's Italian and he plies his father with stories about her. Rome listens to his smitten son, chuckling.

He's also redoubling his efforts with his L-DOPA project, which hasn't yet advanced to anything promising. Many kids would have abandoned such a losing pursuit, but it seems Romano has an unusually high threshold for frustration.

Romano decides to do an experiment to try to find the critical amount of L-DOPA needed to increase the rate at which gelatin hardens. He and Andy decide he should document this work on video.

Andy says, "Your good looks and the word 'L-DOPA' are not enough."

"Should it be dramatic and inspiring?" Romano asks.

"Don't worry about the video for now," says Andy.

"People are going to cry," Romano cracks.

He gets to work filling three plastic test tubes with gelatin. The first tube is gelatin alone. The second and third have 0.1 and 0.25 grams of L-DOPA, respectively. He adds some drops of blue food dye to each, making the goo easier to see.

He's going to place them on a mechanical oscillator, a rectangular plate that when plugged in tilts back and forth, providing a constant and consistent stirring. But within fifteen minutes, all three samples cure to a suspiciously firm consistency. Romano retraces his steps and realizes he added too much gelatin. He has another go, this time drastically reducing the gelatin and then increasing it in small increments. Every time he ups the gelatin, the substance cures faster. This doesn't answer anything, but it's something Romano notes.

When he tells Andy that his concoctions cured rapidly, within fifteen minutes, Andy intervenes. They're both still noodling what, if anything, L-DOPA can bring to Romano's concept.

"We need to fix this," Andy tells Romano. "You're not trying to show it's going to adhere to plastic, you're trying to show it's going to adhere to skin."

Having been around this block with William last year, who also needed a stand-in for skin, Andy tosses the idea of using pig intestines, or sausage casings, to Romano.

"Oh, my god," says Romano. "I bet my grandmother has some." He calls Nanni and sure enough, she's got them in her basement freezer.

The next day, Romano needs to come up with a way to apply his gooey mixtures to the thin-membrane sausage casings—to mimic putting a liquid bandage on skin. Laying out the skin and pouring his goo on top won't do. Since the test tubes are about the diameter of a sausage, Andy and Romano think if they can just cut a window out of the test tube and then slide the sausage casing over the tube, that would do. But the test tubes are made of very sturdy plastic.

After school, they head to the school's woodshop and use a band saw to cut three-inch-long windows into the sides of the test tubes.

Back in the lab, Romano lays out his tubes and begins the job of sliding the casings over them. The skin fits over the windows they cut—and that's where the gelatin and L-DOPA mixture will come into direct contact with the casings.

As he works, it looks like he's fitting a condom over the test tube. Fortunately, this does not appear to have occurred to him. He's also not remotely grossed out by having to fiddle with pig gut linings. In fact, it seems to have rejuvenated him.

Once the modified test tubes are fitted with the pigskin, he fills them with the gelatin concoctions, again using one that's gelatin only and the other two with different amounts of L-DOPA. He tapes them onto the surface of the oscillator and flicks it on. It makes a kind of grinding motor-like sound. This time, he trains an iPad on the setup and hits Record.

He's got a Cheshire Cat grin on his face.

The next day, Romano walks into the lab and runs over to the iPad. He stops the video and hits Play. There on the screen is a close-up view of the swishing test tubes. But he fast-forwards and the screen goes black.

Uh oh.

"Mr. B.," says Romano. "The screen is black for hours."

"Oh, the motion sensor," says Andy, wincing.

Annoyingly, the lab's overhead lights are on a timed motion sensor. If no one moves for about fifteen minutes, the room goes black.

In fact, the only time during the recording that the lights come back on is when the janitor comes in to clean the room. He leaves and then within fifteen minutes, the screen goes black again.

Andy doesn't miss a beat. He walks over to a cabinet and fishes out a desk lamp. It's times like these when you get the feeling you could just start calling out random items and Andy could open a drawer or a cabinet and produce them.

"Set this up and you won't have the light and timing issue," he says. With the desk lamp as a constant light source, it won't matter if the lab's large overhead lights go dark.

So Romano makes new samples, trains the lamp on them, resets the iPad, and turns on the oscillator.

The next day, much to Romano's shock, one of the test tubes is basically empty. The lamp was too close to the tube, so its heat essentially cooked the goo, causing it to evaporate.

"You've got to be kidding me," says Romano in a rare moment of frustration.

Ah, science.

He decides to try one more thing. He prepares fresh samples and sets the iPad to time-lapse mode. In his previous videos, the iPad eventually ran out of space. He thinks that time-lapse will capture more footage in a shorter period of time.

Wrong. The following day, he sighs and rolls his eyes when he watches the video. The time-lapse mode, as you might know, dramatically speeds things up. So now the gentle back and forth of the oscillator appears like a seesaw in a hurricane, violently tipping back and forth.

"It's so fast, I can't tell if it's liquid or solid," he says.

At this point, it's safe to say that the great video experiments are a bust. A solid fail. Not going to be syndicated. To Romano's credit, he sees the humor in what can feel only like cruelty.

"I was all excited, I was talking with Bruce Lee, like, this is possible. I put it on the oscillator. The lights go out. It's like, boom, fuck you,

you suck. Then we have a light source problem. Screw you, still no, you suck. After four times, it's like, ha ha, fuck you. It's like the L-DOPA gods being, like, screw you," he says.

Romano takes a deep breath. He focuses and has a zen realization. He now believes that with all of the mistakes and technical difficulties, he has learned so much that he can now perform the perfect experiment, accounting for all his errors. He spends some time carefully curating his fourth attempt.

The next day, when he comes into the lab, he heads straight to his setup. He turns the oscillator off and examines his test tubes. Two of them have cured to a firm, gelatinous state. He picks up the iPad and starts scrolling through the video. Sure enough, several hours in, the gooey mixture isn't sliding back and forth along the bottom of the test tube. It's curing, getting more firm. It's the test tube containing the 0.25 grams of L-DOPA. He calculates the total time and it's about eighteen hours. According to the video, the gelatin-only sample took longer.

So does this mean the L-DOPA finally did something? Did it make the gelatin mixture cure faster? It seems that way. But here's the problem, which Romano figures out immediately. Remember when he accidentally used an excessive amount of gelatin and it cured rapidly, within fifteen minutes? And then he started with a small amount of gelatin and added it in greater increments? He found that as he increased the gelatin, the mixture cured faster.

When he assembles his midterm science research presentation in slide show format, this is what he writes:

Although promising, a question still remains: Why add L-DOPA to cure gelatin when adding more gelatin has the same effect? This unanswered question is a reason I am looking into collagen.

Collagen? Oh, boy.

Sophia and Andy enjoy a quiet moment in the lab.

Romano working on the antibiotic component of his liquid bandage.

Ethan, who outgrew Greenwich High long before he graduated, works on his carbon capture technology in a lab at Yale.

The many worlds of William: outside GHS, where in four years he took twenty AP exams; with his sister, Verna, at his final home cross-country meet in October 2016; playing the mellophone in the pep band at homecoming; with his prom date, Lauren Low. Lauren and William met on the competitive science fair circuit.

The day he was crowned homecoming prince, Romano posted a shot on Instagram of himself smooching his true love, Nanni.

Olivia Hallisey with her father, Bill, following her presentation about her diagnostic test for Ebola, at the STIR healthcare conference in December 2016.

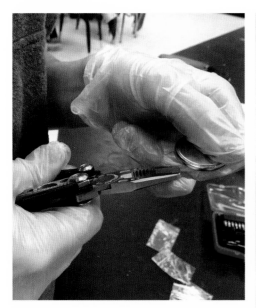

"Danny" works on his portobello mushroom battery. Here he is trying to pry open a battery casing.

Sophia at work on her Lyme disease project.

Tommasina, Andy, and their daughter, Sofia, in September 2014, when Sofia won a ten-thousand-dollar Davidson Fellows Scholarship for her science project.

The Cuteness, the Bramantes' beloved bichon frise, whose passing in the fall of 2016 was one of several low moments Andy dealt with at the beginning of the school year.

While visiting Capitol Hill during the Regeneron finalists' week, Ethan and Derek Woo grab a photo with Senator John McCain.

Michelle Xiong created a water filter that was activated by light.

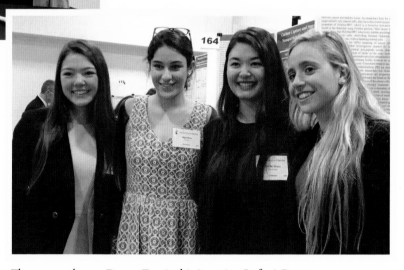

The power players. Devyn Zaminski, Agustina Stefani, Dante Grace Minichetti, and Olivia Hallisey at the Connecticut STEM Fair in February 2017.

Ethan doing some media at the Regeneron STS public viewing day.

Proud papa Gregg Chow with Sophia at the 2017 ISWEEEP competition in Houston, as Andy and Rahul Subramaniam look on.

Ethan, Andy, Shobhita Sundaram, and William at the Connecticut Academy of Science and Engineering dinner, where Ethan delivered a speech.

Romano and Sophia at the Burning Tree Country Club pre-prom party. They each went to the junior prom with other dates, but stopped to pose for a photo together.

Andy and the science research seniors at the prom. (Note: The blurred face is "Danny.")

Sweet! Ethan savoring dessert with fellow Regeneron finalists.

Greenwich meets the Bronx. Romano, Rahul Subramaniam, Alex Kosyakov, and assorted science research kids at Yankee Stadium.

They made it! After three intense years in the lab, Andy and William pose for a graduation photo.

32

Danny

Perhaps it's the new year—or maybe it was all that recharging time during the Slate family's annual holiday trip to Saint John—but Danny has returned to the lab with what appears to be a renewed interest in his mushroom battery.

One Friday after school in early January, he walks into the lab, which is crowded with kids working to get to CT STEM. An after-hours Danny sighting in the lab is like spotting an exotic animal in the wild.

"Did you just wake up out of a coma?" asks Andy.

"New year, new strategy for 2017," Danny replies.

"Good luck with your quest," says Andy. In the walk-up to the year's first fair, he's in crunch mode and he devotes all his time and energy to the kids headed to the competition. Danny's on his own, which normally wouldn't trouble Andy. But as Danny's been away from the bench for a couple of months, he's a touch out of practice.

He needs to construct the batteries in a glove box, one of the encased glass work spaces. But first he needs to remove all the oxygen from the glove box and replace it with nitrogen because oxygen will oxidize (rust) the lithium in the battery. The lab's nitrogen tanks stand about five feet tall. They look like small missiles.

"Tuesday after school, I accidentally used up the nitrogen tank," he says. He's not sure what happened, but in his zeal to make sure the lithium wasn't exposed to oxygen, somehow the tank got left on—for three hours.

As Danny explains this, Wesley Heim, who is busy working on his own project—he's trying to embed carbon nanotubes into a performance fabric that could radiate body heat when worn—says, "Oh, my god." Wesley is a very methodical worker and hearing about these gaffes makes his head hurt.

Since Danny was unable to persuade Panasonic to sell him empty battery shell casings into which he could have easily slipped his mushroom-coated silicon wafers, he's resorted to an alternative.

"Christo had a bunch of extra casings and he's nice enough to let me use them," says Danny. "I cut [the wafer] up and put it in the casings. The voltmeter will be used to see if it's working." A voltmeter does what its name implies. When you hook it up to an energy source, it displays a reading of the output, or voltage.

But first, Danny needs to run down the lab's coin battery charger, which he'll use to charge his battery ahead of voltage testing. Instead he starts chatting with Henry Dowling, who's also made a rare after-school appearance, and gets sidetracked.

A week later, on Friday, January 13, the mushroom battery still hasn't been charged and tested.

Danny walks into the lab with a look of distress on his face. Madeleine Zhou says, "Danny, what are you doing here?"

"I finished a math test early and I feel like I died inside," he says. "It wasn't a great math test."

Madeleine narrows her eyes at Danny and says, "I think you got a haircut last week and it looks like you need another one."

Later, after he seems to have recovered a bit from the math test trauma, Danny decides he's going to charge the battery. He has finally managed to locate the coin battery charger and has it in hand.

"This is what Christo used to charge his. I hope it will work; I don't know if it will," he says, placing his battery into the charger and plugging the charger into one of the outlets that sits atop the lab tables.

Wesley is experiencing some issues with his project and Danny offers up some thoughts. He's encouraging Wesley to look at molarity, a measure of the concentration of a solution.

"I don't think moles apply to carbon nanotubes," says Wesley.

"It should. The chemist in me likes molarity," Danny says.

"The chemist in you . . . ," says Wesley, shaking his head.

Danny then launches into a long, eloquent monologue about the use of dopamine injections to treat Parkinson's disease, which he somehow eventually links to what Wesley's working on.

"How do you know all this?" says Wesley, equal parts impressed and baffled by Danny's sudden display of random scientific knowledge.

"I took a neuroscience class," replies Danny. "And I don't have a life."

There is, however, one very big and exciting thing looming in Danny's life: his road test to get his driver's license, scheduled for March 24. The number of family members who share in this excitement is zero, but his parents recognize it's something that must be done. For his part, as a senior, Danny doesn't seem to love being chauffeured around by the Slate family's nanny, though he remains grateful that she's the pizza runner.

Should Danny pass the test, he has a car waiting for him. His grandparents have gifted him their 2000 Lexus SUV. It's sitting on the castle grounds, in need of a licensed driver.

Danny's driving has been the source of considerable mockery, though Danny maintains he's good and capable behind the wheel. But even with his eagerness to drive, according to Danny, his 2017 priorities are as follows: "Finish the research project, that's gotta be number one. Go out more, that's up there with number one. Oh, and for my senior project, we're making an internal combustion engine . . ."

On Friday, January 27, Danny's standing in the lab, holding the voltmeter, which is hooked up to his battery.

"Mr. B., it works!" he says after reading the meter, which shows that his battery is producing an excellent amount of energy. Danny's stoked.

"Just make sure it wasn't the control or something," Andy replies, half joking.

"No, it can't be," says Danny. But then his jubilant expression starts to melt. He looks at the battery and seeing its perfect, untouched seams, realizes it is the control: a store-bought battery. This isn't the one he built. Ugh. No biggie, he thinks. Annoying, but whatever. He just needs to charge up the one he made, which will take a day.

He's going to start right now. Danny's looking around the room trying to locate the small clear-topped rectangular case where he keeps his battery. He doesn't see it. He's sure he left it on a cart. Or maybe it was the top shelf. Or the glove box. *Where's the battery?*

He starts pacing frantically around the lab, clutching a yardstick in one hand and nervously tapping it into the palm of his other.

"This is the kind of thing that would only happen to me," he mutters. "This is what I'm good for."

The jolly mood of yesterday has vanished, along with the battery.

"I still don't get how it could be gone," he says.

Lily McKenna looks up and says, "God works in mysterious ways."

"Way to pile on, Lily," says Danny, unamused. Lily isn't actually a science research student, though she does make regular appearances in the class.

When asked who she is, Andy replied, "Oh, that's Lily. She's a groupie." Lily hangs with the science research kids and shares some of their classes. She just likes to chill out in the lab in her free time and Andy doesn't mind because she's sort of become part of the fabric of the class, a fun, funny supporting player.

Danny's trying to stitch together the chronology to figure out how that sneaky motherfucker of a battery made off and double-crossed him.

"January tenth, I made the battery," he says. "It was last seen on January thirteenth." Which is exactly two weeks ago. Danny starts flinging open cabinets and drawers, looking in, around, and on top of instruments. The lab is unfortunately a vast space with endless nooks and crannies—any number of which could be concealing his quarter-sized battery. Imagine trying to find a single beer bottle cap in a two-car garage stacked with boxes and bins galore.

Andy is bewildered.

"You LOST it? Where the hell did you put it?" he says.

"If I knew *that* . . . ," says Danny.

"Did you set it down somewhere? Have you looked in all the drawers?" asks Andy, incredulous that after taking four months to assemble, the battery grew legs and walked off.

Andy watches Danny doggedly search the lab, letting him suffer a bit as penance for being so knuckleheaded, and then gets involved. He's now picking through lab rubble, weeding through boxes. No sign of the mushroom battery.

"I guess I'll just make another one and immediately go to testing," Danny says. "Okay, here's my course of action. Monday, what I do is break up silicon wafer and put it in a battery. I'd have to make more anyway."

Then he remembers other obligations.

"Monday is math team. I'll probably stay after school on Tuesday. Tuesday, I'll blow up the glove box, pour in the nitrogen. Wednesday during class, I will test the battery. Thursday, I'll come back and see if it all works," he goads himself during this self-flagellating pep talk.

Danny knows how completely farcical this appears. It's as if the universe is giving him the finger. *You dragged your feet, Slate. You didn't get your shit together. Let's see what happens when we take away your little battery.*

"It feels like it's disappeared off the face of the earth. What could have happened to it? That's what just doesn't follow with me." He's in a verbal whirlpool now, circling the drain. "I never took it out of this room. It doesn't compute."

33

Ethan

B ack in November 2014, Ethan was working in Andy's lab when a strange thing happened. He was trying to separate ethanol from water—something that requires a lot of energy. Ethan had some vague ideas on how to improve the process.

He added ammonium bicarbonate to the ethanol-water solution and much to Ethan's surprise, a bunch of bubbles starting forming. "I had an ammonia sensor and CO_2 sensor and the CO_2 sensor went crazy," he says.

The addition of the ammonium bicarbonate to the solution had a curious effect: it was capturing carbon dioxide. He started adding other solvents to see if they, too, would capture CO_2. Many of them did. This was one of those fortuitous moments in science that happen despite all the painstaking efforts to control and micromanage every step. He knew he was onto something and so the following year, while at Yale, he started devoting his time and energy to carbon capture.

About a year later, in December 2015, he heard that someone from the Carbon XPRIZE was coming to campus to explain and promote the competition. XPRIZEs are awarded in a variety of fields, every-thing from carbon capture to oceanic discovery to adult literacy. The purses are in the millions. In the case of carbon capture, teams are judged on two criteria: the amount of carbon dioxide converted from power plant waste and the net financial value of the products created from the captured CO_2.

Listening to the talk, Ethan grew increasingly excited. He was feel-ing confident about his work, and he'd shown he's a master competi-tor on the science fair circuit. After the discussion, he bounded up to the guy, shared his technology, and started thinking seriously about entering the Carbon XPRIZE. If he was one of the winners, he could take home $7.5 million. But even entering had substantial hurdles.

For starters, he'd need to form a company, an actual business en-tity, in order to compete. In February 2016, he launched Innovator Energy. He initially brought on the two Ph.D. candidates from Pro-fessor Menachem Elimelech's lab as members of both his XPRIZE team and his budding company. But the relationship became strained. Ethan felt he was doing all the work, yet the two researchers wanted a sizable equity stake in the company. Eventually, Ethan decided to sever ties with them and plow forward on his own. He retained only Elimelech as an advisor and official member of the Carbon XPRIZE team.

Ethan officially applied to the Carbon XPRIZE in July 2016 with his novel method. Here's a description from his company website:

> The process introduced in our ACS peer-reviewed paper requires 75% less energy than existing CO_2 capture processes. Our confidential CO_2 capture technology, CO_2 Evolution™, is the first entirely powered by discarded, abundant low temperature waste heat with a temperature as low as 28 degrees Celsius and has 90% lower operating cost than existing CO_2 capture processes.

In October 2016, Ethan is named one of twenty-seven semifinal-ists. All of the other twenty-six start-ups, some of which grew out of academia, are made up of at least two people, but some have as many as sixteen. And in many cases, the team members are old enough to be Ethan's parents. While Ethan lists Elimelech as an advisor, at this point, he works entirely solo. He got a utility patent for his work, funded in part from his various science fair winnings over the years.

According to the Carbon XPRIZE criteria, the semifinalists are required to "demonstrate technologies in a controlled environment (such as a laboratory), using a simulated power plant flue gas stream. Teams must meet minimum requirements and will be scored on how much CO_2 they convert and the net value of their products."

This means that teams must build a prototype of their carbon capture technology and meet certain performance thresholds to advance to the final judging stage.

Where on earth is a seventeen-year-old going to come up with the money to build a prototype of a carbon capture unit intended to be attached to a power plant?

By Ethan's own estimate, his technology will cost about $300,000 to construct—which is way more than he has in his personal piggy bank.

The Noveks consider investors—Ethan has never wanted for interest in his work—but that would mean giving up equity, possibly losing control, and, of course, working under a ton of pressure to keep stakeholders happy.

He also needs to contract with a facility that can build and test his carbon capture unit. "I had two routes," says Ethan. "I could build it at Yale," he says, but no one he's working with has any experience building anything like this.

Or he could try an independent, nonprofit research and development facility in San Antonio, Texas, called the Southwest Research Institute. SwRI works with industry and governments to construct massive projects. It's done extensive work with NASA. While it's never contracted with a seventeen-year-old before, that's not to say it won't. Ethan decides to submit a work proposal to SwRI.

"I was expecting the price to be insane," he says. "But they realize the novelty of the work."

In November, SwRI comes back to Ethan with a proposal, the terms of which are confidential. The numbers are all relative, but the figure they quoted the Noveks was workable.

Bonnie and Keith decide that, for now, they will personally fund the cost of Ethan's prototype. Neither has any scientific background. Keith is a management consultant at one of the Big Four accounting firms. Bonnie has worked as a realtor, but these days, she tends to her sons. Ethan's older brother, Chad, is a student at Northwestern. He's also got the Novek entrepreneur gene. He operated a highly successful tutoring business while in high school.

The Noveks don't spend a lot of time overanalyzing or agonizing about their decision to back Ethan's work. For now, he's producing such unbelievably strong results, they don't see the outlay as a high-risk proposition. Ethan is pleased because he and his family retain 100 percent equity in Innovator Energy.

It's critical to note that everything that makes Ethan unique is propelled forward by the great fortune of coming from a family of both financial means and unwavering support. There aren't many kids like Ethan, but there also aren't many families who'd be willing and able to allow their child to chart his course in this manner—and to wager this kind of money.

Around the same time, Ethan gets an intriguing email from a technology company in Cupertino, California, called Apple. Apple found Ethan through the NRG COSIA Carbon XPRIZE site, which links to the website for Innovator Energy. Someone from the company's materials division invites Ethan to Apple's headquarters to deliver a presentation on his carbon capture technology.

Since he's too young to rent a car or check into a hotel on his own, Bonnie accompanies him, as she does on all of his trips that require such logistics. "She removes barriers," Ethan says.

So it is that one sunny day in November 2016, Bonnie drives Ethan to Apple's headquarters, drops him off, and then goes and gets a manicure while her son does his thing. She never thinks about inserting herself at such meetings because to do so would be to undermine his credibility and draw further attention to his age.

Ethan says, "When I get really, really excited about something, I can't eat. I was just so excited. I knew this was an incredible opportunity."

He stands before a roomful of engineers and material scientists and explains his technology. Well into the discussion, some of the folks start asking about his Ph.D. work.

"They just assumed I was a master's or Ph.D. student," he says.

It is at this point that he reveals he's a seventeen-year-old high school student. He declines to say that his mother had dropped him off that day because he can't rent a car.

"Apple is one of the more accepting places," he says. "Unlike other incidents, the discussion didn't immediately go to me as a kid. It ended up being all about the technology. It didn't distance me from the conversation."

Apple explains that their interest is twofold. When Ethan's work is at a later stage, they may have interest in investing in this area, and also, given its extensive manufacturing around the globe, it has a vested interest in reducing its carbon footprint. The whole experience is exhilarating.

The next month, December, also turns out to be a stellar one for Ethan. He signs a contract with SwRI. He gets into Yale, early admission. While it's clear Ethan was never meant for high school or vice versa, the question of college remains an open one. Sometimes the family has dipped its toe into the idea that Ethan might not need college either, though that feels scarier and college is certainly not as disposable as high school. After all, college is where you make lifelong friends, often fall in love, learn to play beer pong, all that stuff.

Ethan's surprisingly stoked when he gets his acceptance, even posting on Facebook, "Yale Class of 2021!!" to lots of kudos from his friends. Of course, no one ever thought there was a chance he wouldn't get in. It's refreshing that even after he's published a peer-reviewed paper, worked at Yale, advanced in the Carbon XPRIZE, and presented at Apple, college acceptance still sparks some excitement.

For now, he realizes he needs to move to San Antonio to oversee the construction of his carbon capture unit. So, much like they did in New Haven, Ethan and Bonnie head to Texas to help him set up what will become his new life.

34

Sophia

One day in late January, Sophia walks into the lab carrying a navy blue Zara shopping bag.

"What's that?" asks Charlotte Hallisey.

"It's the Model Magic for after school," says Sophia.

Senior Margaret Cirino comes over, and as Sophia shows her the contents of the bag, Margaret beams with excitement.

"Oh, this is great," she says. Sophia smiles, too, because it's always nice to get Margaret's approval. As one of the rare four-year veterans of science research, as well as someone who enjoyed a winning run on the competition circuit, Margaret commands a certain respect in the class. She's an old soul with a natural air of poise and know-how. You likely could have asked her when she was in fifth grade to change a roadside flat or bake a chocolate soufflé. Margaret is someone you'd want around during any kind of catastrophe because of her ability to make sharp, rational decisions and remain calm.

In one of the more altruistic student endeavors, Margaret started STEM Explorers, a weekly after-school science club for fifth graders at Cos Cob School, one of Greenwich's eleven elementary schools.

She and a rotating cast of Andy's science research students trek over to the school on Thursday afternoons from three to four thirty to lead the club.

In keeping with Margaret's penchant for organization, purpose, and thought, STEM Explorers isn't a free-for-all where kids throw pipe cleaners at one another or play videogames or stare into space. Each eleven-week rotation has a theme and the time is spent learning things related to the chosen topic. During the last half hour of each session, the kids work on individual science projects of their choosing and design—miniature Margarets and Sophias in the making.

The current theme is underwater life and today is Sophia's day to lead the club. The kids run into the school's Maker Space, the ingenious creation of the Cos Cob School librarian, Nancy Shwartz, a study in human warmth and smarts with the added benefit of looking like a middle-aged version of Ms. Frizzle, the science teacher in the children's book series *The Magic School Bus*.

The Maker Space is unlike anything you've ever seen, a Willy Wonka–like fever dream of gizmos, gadgets, parts, and ideas. Thousands of LEGO pieces are neatly arranged by color in clear bins. There's a green screen to shoot video, including tripods specially designed for iPads ("to get rid of the *Blair Witch* effect," says Mrs. Shwartz, mimicking young kids trying to hold an iPad still while filming); a wall of littleBits circuit pieces; and perfectly organized boxes of string and felt and colorful pom-poms and a zillion other things. The motto of the room is painted on the wall: Create without fear.

Mrs. Shwartz built the room as a place where kids can try to make just about anything—and where the stakes have nothing to do with grades or traditional performance metrics. "All day, we're telling the kids, do this, read this, use this—and if you don't, you fail. They need a space where it's okay to fail. It's okay for you to fail here. There's so much pressure in Greenwich on these kids to be perfect."

And so she was thrilled to host STEM Explorers. To Mrs. Shwartz, another added benefit is that the club is led by girls. "It's important for [the younger kids] to see *girls* can do this."

Her point is a salient one. According to the National Girls Collaborative Project, which seeks to encourage girls to pursue STEM careers, during the K–12 years, male-female participation and achievement in STEM courses is equal, with the exception of computer science and engineering. Disparities start to emerge in college. Though women receive more than half of undergraduate degrees in biological sciences, the numbers shrink to 39 percent in physical sciences and 19.3 percent in engineering. (The figures are most bleak for minority women: in 2012, they constituted only 4.1 percent of doctorate recipients in STEM.) Given the decline in college, it's not surprising that even though women account for more than half of the college-educated workforce, they constitute only 29 percent of the science and engineering workforce.

Today, Margaret and Sophia are joined by Dante Grace and Charlotte. They all have a nice touch with the younger kids, speaking to them encouragingly but unafraid to quell the inevitable rowdiness.

In the main library, a delightfully inviting room softly lit with lamps instead of fluorescent lights and appointed with lots of cozy reading nooks, Sophia shows them a short clip from a 2007 TED talk about bioluminescence in sea creatures. The video has a mesmeric effect and the little ones watch with great interest. Once the video wraps, Sophia and Margaret tell all the kids to gather in the Maker Space. The four boys and four girls run to the room.

Sophia asks them a few questions about bioluminescence, which they eagerly answer.

"Now you all have Model Magic on your tables. You have to build a sea creature and make it light up," Sophia tells them, to squeals of "Ooooh" and "All right!" The kids pick up the moldable, claylike substance and start playing with it.

Mrs. Shwartz is here as the adult in the room and keeps general order when the need arises but lets Margaret and her team run the show. She looks on with pride while watching the older girls take command. Each of Mr. B.'s protégés sits at a table with the younger ones, a ratio of one high school girl to two fifth graders.

One of the kids is a motor mouth who seems a tad impulsive. There doesn't seem to be anything serious or worrisome at play, he's just a fast-talking, loud kid who requires some work to rein in.

Margaret walks over to the table with Motor Mouth.

"Have you guys decided on your sea creatures?"

"Yeah. I'm gonna do an octopus," says Motor Mouth.

"That's a great one!" says Margaret, all cheer and encouragement.

"Yeah, all the testicles," he replies.

Um.

The realization of what's been said lands with great force. In terms of awkwardness, it's right up there with audible flatulence. The look on Motor Mouth's face clearly indicates this was a grave mistake, not something said for laughs. Oh, the difference one consonant can make.

"I just said testicles. I meant tentacles. Why did I say testicles?" says Motor Mouth, frantically trying to motor-mouth the embarrassment away. In an award-winning moment of utter grace, Margaret doesn't startle, flinch, or even make a face. She just glides away to another table, which truly is the only acceptable move in this situation. After all, she's seventeen years old, therefore hardly immune to deeply uncomfortable situations.

Luckily, the room is noisy enough that the slipup is heard only by Motor Mouth, Margaret, and Motor Mouth's partner, who seems somewhat traumatized. The moment could have quickly evaporated, but a full three minutes later, the recovery effort is ongoing. Motor Mouth can't shake it off.

"I said TESTICLES when I meant TENTACLES," he says to his partner, who at this point just really needs Motor Mouth to stop talking.

Meanwhile, Sophia is helping two girls who are making a museum-worthy octopus, one that, it must be said, has no testicles. The kids tinker with the light-up circuits, debating whether to embed them beneath the modeling clay to make their creatures light up from within or to add them on the outside, for a more decorative look.

Sophia calls them all together to share their projects and the kids describe their choices. There's a nice array of colorful creatures, some real and some imagined.

There's a bit more talk about bioluminescence and then the kids are free to work on their independent projects. Some are still in the planning phases.

Sophia powwows with two precocious girls. As she's talking to them, her face is like a shifting kaleidoscope, turning from smile to curiosity to puzzlement. At some point, she gets up from the table and wants to huddle. She lowers her eyes and talks in a hushed tone.

"They want to add different amounts of salt to water to see how much salt baby shrimp can live in," she says, punctuating this with a look of horror. The girls have designed a threshold experiment that has an undeniable whiff of animal cruelty.

"They're going to kill the baby shrimp," says Sophia, a look of disbelief frozen on her face.

But since she doesn't want to also kill their learning experience, Sophia isn't sure whether it's within her purview to file a petition with the girls to save the shrimp.

Charlotte intervenes and asks the girls if they are concerned about the potential shrimpicide.

"No," they say, nonchalantly.

"Okay," says Charlotte and jokingly adds, "We're not so much into the animal welfare around here."

Sophia's feeling a bit less dragged down by her Lyme disease, less lethargic, so she's slowly doing more activities. In addition to STEM Explorers, she has started taking tennis lessons twice a week after school at Innis Arden, the family's country club.

She admittedly has a hard time being medication compliant—meaning that since she was diagnosed, she's never been able to complete a full six-week course of antibiotics. The side effects make her too sick, and she's just not that comfortable taking giant pills twice a day. She tends to take them when she's feeling especially bad. But overall, these days, she says she's got more energy, thus her decision to take tennis lessons with the goal of trying out for the Cardinals team in the spring.

Sophia's project is going well, but there's one thing that's a significant source of worry. In order for the biofilm to be viewed under the SEM, it needs to undergo "fixation," a chemical preparation that has to be done by professionals. Mount Sinai generously did this for her last year, but given that it's hardly their mission or priority, it took a long time. She'd much prefer some quantitative measure that doesn't rely on the SEM, the schlep to the city, the inability to control the timing.

So one night at home, she opens her laptop and starts to search for alternatives. She plugs in terms like "measuring biofilm" and then "measuring biofilm growth thickness" and then "measuring biofilm growth thickness with a microscope." She scrolls through the results and finds a paper on PubMed.gov, a massive trove of scientific papers. The title of the paper is "Optical Reflectance Assay for the Detection of Biofilm Formation." She clicks on the abstract, which says researchers have identified an "inexpensive and nondestructive" method to measure biofilm growth using a reflectance spectrometer, an instrument that measures how much light is reflected from, say, the surface of a biofilm before and after drugs are applied.

The realization sinks in. She doesn't want to get ahead of herself, but Sophia's gut tells her she's landed on a game-changing possibility. She prints out the paper, staples it, and tucks it into her backpack.

The next day, room 932 has the vibe of an industrious factory. With just nine days to go before CT STEM, kids are working with intensity and purpose. Andy's posted the names of the ten kids who will be going, those who did an early push with their projects and are now either done or putting finishing touches on their work. It's ultimately Andy's call who goes, and he bases his decisions on who's actually fair ready. He's bringing a block of six girls that includes Olivia, Devyn, Dante Grace, Shobhi, Agustina, and Sophia. The boys

are seniors Connor Li and Adam Roitman and two first-timers, Noni
Lopez and Steven Ma.

Andy's steadily making his way from kid to kid, checking in.

Sophia approaches him. "Hi, Mr. B."

"Hello, Ms. Chow! Yes? What can I do for you?"

"I want to show you this paper I found. I was looking for alternative ways to measure the biofilm so we don't have to use the SEM at
Mount Sinai," she says.

Andy takes the paper and gives it a quick read.

"Wow. Look at this," says Andy, intrigued. "If I'm understanding
this correctly, we can load your samples into the spectrometer and
measure the reflectance of the biofilm with your various drug combinations."

A smile spreads across Sophia's face, and then Andy's. What are
the chances that the lab just happens to house a reflectance spectrometer, one of Andy's signature instruments from his industry
days? Spectrometers are like old friends to him. He knows their every
quirk and tic.

"All right," he says. "I don't see why this won't work. Let's do it.
Great job finding this paper, Sophia."

Sophia is thrilled. She'll be able to fully execute the back end of her
project on her own. This takes on a greater meaning for her than it
might for other kids. For the three years her health has been compro-
mised by Lyme disease, she's had a feeling that certain aspects of her
life were beyond her control. It's been hugely dispiriting not only to
feel so foggy and sick but to not be able to quash her illness. But now
Sophia can get the project over the finish line right here in room 932,
and she's fully in charge. The remedy she found is enterprising and
will also give her a new skill, using a spectrometer.

So today she has a different feeling. It's one of elation, of owning
her work, of getting ahead. It's not something that comes often in the
maelstrom of school, where it can feel like the work at hand consists
of cramming information into her head only to spit it back in test
form. But this is hers. And it feels amazing.

35

Andy

It's February 1, three days before the Connecticut STEM Fair. Today the final block of the day is William's, the boisterous, overstuffed research class that nudges Andy ever closer to the cliff of sanity. He parks himself with his laptop at the table closest to the door, as if readying for an escape, hoping to go over some finishing touches with Shobhi. There's a binder on the table with color printouts of the fair posters in neat clear plastic sleeves.

Emily Philippides opens it up and sees Olivia's poster for her saliva test to detect Lyme disease.

"Ah, in the presence of greatness," Emily says, a starry look in her eye.

Within minutes, the lab is buzzing with activity, not so much with the kids Andy's taking to the fair, but with the ones he isn't. There's something about the appearance of the posters that lights the proverbial match under butts, he says. He's pleased to see the flurry.

The days of a library-like atmosphere where the kids quietly research and read are over. They now rush into the lab with urgency, chuck their backpacks anywhere they can find a spot, and push up their sleeves, their faces taut with purpose. Cabinet doors are flinging open as they rummage for beakers, flasks, and pipettes. There's a staccato of latex gloves being snapped on, and if there was ever any self-consciousness about looking like a mosquito while wearing protective goggles, it's gone.

The issue with all this activity is that every single kid needs something from Andy to keep things moving. Joe Konno needs to make a water bath to heat a solution, Noah Kim isn't sure how to pipette half milliliters, Verna wants some input about next steps with her iron nanoparticles. And all the while, Steven Ma is waiting for Andy's undivided attention so he can test-drive his oral presentation before Saturday's fair.

Like a flash flood without warning, there's a collective of kids hovering, lining up, pacing, and circling around Mr. B., waiting to pounce to get his attention. The second he finishes a sentence, there's another kid firing. He's trying to be zen, nursing an orange Tootsie Pop for comfort.

Steven Ma approaches. Steven's default look can be interpreted as glum or pensive. But either way, it's that of someone perpetually possessed of very big thoughts. Andy looks up for a millisecond from his laptop.

"Hello, Steven. What sad tale do you have to tell me today?" asks Andy.

But before Steven can ask his question, Noah says, "When you have a minute, can we talk?"

"Yeah, see you next week," says Andy.

Steven asks, "With the abstracts, do you have them with you? And are you bringing the posters to the fair or are we?"

"No, I'm not bringing ten posters," says Andy. "This is your deal, man."

Verna steps in. "Mr. B.?"

Andy says to no one, "Man, I'm going to get a clicker," and then mimes using a people counter, his thumb doing a rapid flicker. Verna needs some chemical that Noah's using.

"Are you nervous or excited? Are you engineering or life sciences?" Andy asks Shobhi, checking on which category her project is in. In a rare moment, she's rather quiet, as she's focused on something on her laptop.

"I'd better be engineering or I think I'm screwed," she says, now furiously trying to confirm her category.

Andy shows Shobhi the judging sheet.

"This is the score sheet. You want to make one or two references to previous research," he says. "These people are forced to go down this list. Whether they like it or not they're going to assign you a number from one to fifteen."

"In terms of recent sources, should I say when the research was conducted?" asks Shobhi. All the kids at the fair must have their lab notebooks, which document their work, and a binder with their resources, prior work, and printed abstracts describing their project.

"Is it okay to annotate?" she asks. "Do I need printouts of all the code I used?"

"I assume you're going to have some interactive stuff," says Andy.

Emily takes a lollipop out of her bag and puts it in her mouth.

"You inspired me, Mr. B.," she says, smiling.

Noah's now all goggled, gloved, and holding a tub of some sort. "Mr. B., stupid question, but how do I open this?"

Andy eyeballs it and says, "Take off the tape. It's wax. Make sure you do this under the hood, full regalia."

Within a nanosecond, Collin Marino says, "Mr. B., I'm going to need to incubate something for two hours at thirty-seven degrees Celsius."

"Get here early," Andy says, laughing.

The repeated incantation of "Mr. B." from demanding teenagers feels like an enhanced interrogation technique, a mild form of torture.

"Mr. B., this is what I was going to show them," says Shobhi, steering her laptop to Andy. "This is a pancreatic cancer one."

Shobhi takes the data from a patient blood sample and runs it through her algorithm, which can indicate with 80 percent accuracy whether someone has pancreatic cancer. In the example she's showing Andy, the screen says "Malignant."

"I would spend some time explaining what [the algorithm] did," advises Andy. As he says this, he notices someone has placed a glass bottle perilously close to the edge of the table's corner. He tells Emily to slide it back.

Collin returns, wanting the green light to heat his chemical concoction.

"Throw your stuff in," says Andy.

"Do you think you can take it out in two hours?" asks Collin, somewhat reluctantly.

"Nooooo," says Andy. "Where are you going to be?"

"At my dad's," says Collin.

"Write me a note. I'll be here with the STEM kids," Andy tells him, meaning he'll be staying after school for further prep.

Emily's looking at the fleshy pink middle of her lollipop, from which she's taken a good chomp. All of a sudden, with the same urgency as if she's accidentally drunk antifreeze, her voice shoots up a few octaves, and she says, "Mr. B., is this gum? Am I supposed to eat this? Can you eat this?"

She's panicked. The girls at the table lean in, have a look, and render the verdict that the middle is bubble gum.

"Ugh, I swallowed it! I chewed it and swallowed it."

"Now you're going to fart bubbles," Andy deadpans.

Emily examines the center of the lollipop and says, "Oh, now I see it's gum."

Verna is looking at Andy with animal-shelter eyes but saying nothing.

"What's the matter, Verna? Eye contact with a Yin is lethal."

"It says I need a dispersion."

"You need to put those in solution so they float," he says. "Start with a solid, weigh it. It should redisperse."

Like the Ghost of Christmas Past, Steven has made several return trips to the table, saying nothing, waiting for Andy. Finally, Andy stands up and says, "All right, Steven, let's go across the hall."

They head to the instrument room, a virtual meditation sanctuary compared to the lab. Andy pulls up a chair while Steven props up a version of his poster that's printed on a regular piece of 8½″ × 11″ paper.

Andy gives Steven, who quite possibly was wearing sweatpants in utero, a once-over.

"Before I forget, you have to wear something else. What you're wearing right now—"

"I know. Not good."

"Yeah, no. We'll talk about that later. Okay, Steven. I'm ready for you."

"Many great scientists, such as Newton, Kepler, and Gauss—"

"Bramante's not in there?"

"Sorry, what?"

"Bramante's not in there? I'm kidding. Continue."

Steven then launches into a brilliant but dense explanation of his project.

Steven has spent the past several months on the science of packing a box. Seems simple enough, yes? Using disks as stands-ins for packing materials around items, Steven mathematically proved that a certain configuration of disks was optimal in terms of maximizing stability of the packed item. This uses, in part, jamming or jam-packing to create an ideal packing environment.

Steven begins his presentation by describing the history of packing models and then details the math behind it all. "Originally, Kepler conjectured in 1611 that the densest 3-D sphere packing was the face-centered cubic. Many renowned mathematicians worked on the problem over the years, but it was only solved in 1998 when Thomas Hales used a computer to prove Kepler's original conjecture," he says. You need both an advanced degree and a mental chain saw to cut a path through this information.

Steven's nervousness is seeping from every pore. He doesn't quite know what to do with his limbs. His feet are splayed too far apart, and each time he tries to direct your attention to the paper that's meant to simulate his poster, he pivots and slides his foot toward it, spreading his legs even further. He's now awkwardly low to the ground in a lunge position and has the lightest of jazz hands going. You keep thinking he's going to slide his feet closer together, but no. The stance-hands combo makes him look like he's selling his project on QVC, but his go get 'em posture is at odds with his timid delivery. Whenever he says "jam-packing," he puts the emphasis on "pack," which is weird but strangely penetrating and effective. (You'll find yourself over the next week repeatedly saying "jam-packing" in your head to see where the emphasis should be and find there's no ideal place.) There's so much going on here that it's hard to, well, unpack it all.

It's so bad, it's good.

You keep waiting for Andy to yell "Cut!" But the moment never comes and he doesn't move to reset Steven.

Instead, with zero irony, Andy sits there, nodding encouragingly. He's remarkably relaxed. The look on his face suggests he's watching a Tuscan sunset. He keenly listens but doesn't exhibit any rambunctious reactions. Steven is in need of some emotional scaffolding, and the hearty head nods are exactly what he can handle.

You watch this and cannot help but be moved by the subtle and perfect sweetness unfolding before you. You cannot believe Andy's restraint and the sudden disappearance of all of his wiseass instincts. He is utterly and completely with Steven in this moment.

"Each disk can only move a minimal amount. If you want each disk to move an infinite distance, you have two different criteria which can be satisfied . . . In conclusion, for lattice packings, my main results state that if the angle of lattice packing is 60 degrees, two adjacent disks have to be removed. If the angle is not 60, then you can take away one disk and globally unjam the packing."

The end.

Andy's still nodding.

"In the beginning," he says, "there were a lot of 'we's and 'our's there. I don't know who 'we' and 'our' are. It's you."

"Okay."

"I wasn't getting the gist of the application until you got here," says Andy, pointing to a section of the poster. "Give them a little tickle of what the selling point is going to be, a little floater. The whole point is to design a new way to pack materials. Done. Move on. But at least I have something in my head of what you're looking at. I don't appreciate it for what it is until I start to spin my head around what you're going to do with it."

"Okay. I have to go. Math meet," says Steven, considered to be the school's lead nuke in competitive math.

"Get outta here," says Andy.

As Andy sees it, Steven's got a few hurdles, ranging from wardrobe challenges to finding a way to make packing have some zip. But it's all relative. In many cases, getting to the fair *is* the win. Bringing such a timid kid out of his house and into a public arena, getting him a day without sweatpants so he can deliver a formidable presentation of his work and get credit for a cool idea—for Andy, that's a victory right there.

However, on Friday, the eve of the fair, because he can't help himself, Andy's taking another run at the wardrobe issue. He can see that Steven's nerves are practically tap-dancing and he knows they'll be worse if he shows up looking less polished than the other kids. He could do with the confidence boost that comes from looking a little snappy.

And the message seems to be circling, if not yet landing in, Steven's brain. He's turned up today in khakis.

"What about these?" he says, gesturing down at his ill-fitting, rumpled pants. Despite his not being tall, every single pair of pants he wears are high-waters that hang low at his waist but are flood ready on the bottom. Today is no different.

"Uh, no. They're grody. They look like you slept in them," says

Andy, who slows down, takes a deep breath, and assumes a fireside chat tone, a soothing, instructional voice.

"Here's what you're going to do. You know where Joseph A. Bank is? On the Ave.? I think it's across from Dunkin' Donuts. You know where that is?"

"No."

"Okay, well, you're going to go to Bank and tell them what you're doing and what you need it for. First, get some khakis or trousers. And then you need a jacket. Do you have a jacket?"

"No. What kind of jacket?"

"A sport coat. Navy blue is nice. Get navy blue. Black is too . . ." Andy trails off and then directs his gaze downward. "Shoes? You have shoes?"

"No."

"Okay, those won't do," he says, pointing to Steven's sneakers. The laces are tied into big droopy rabbit ears.

As Andy is walking Steven through the procedure for clothing acquisition, Steven has a look of bewilderment on his face, his eyes wide, his jaw slightly slack. It's as if Andy's handed him a shotgun and told him to walk into the woods, kill his dinner, and cook it over an open fire.

After saying virtually nothing, Steven offers this, "My brother once went to Macy's."

"Macy's! Macy's is great! Go to Macy's!" Andy says, all verbal fireworks. But then he thinks better of it and again tries to nudge Steven to Jos. A. Bank, due to the on-site tailoring.

"Most guys can't just walk in and pluck something off the rack," he says. "You go to Bank and they have someone right there who can throw a seam in for you. Boom, you're done."

Saturday, February 4, is a freezing winter day drenched with bright sunlight, the kind that sunburns your face on the ski slopes. The Connecticut STEM Fair is being held at Darien High School, a whiter, more affluent school than Greenwich High. The fair is for Fairfield County schools, as well as Amity Regional High School, and kids compete with both completed projects and proposals for projects they haven't yet started. Andy's kids bring only the former.

The fair sprawls across two gyms, and all but one of Andy's kids are stationed in one of them. Adam Roitman, who developed a way to neurologically block addiction, is by his lonesome. Adam wears big wire-framed glasses and looks like a cleaner-cut version of Harry Potter. He's sporting a bow tie with a rubber ducky print, a navy blue blazer, the requisite khakis, and nice mushroom-colored suede shoes. He's had his hair cut for the occasion and looks presentation ready.

The rest of the GHS crew are in the other gym, spread out over three rows. Each kid gets table space for his or her posters, models, laptops, binders, and abstracts. There are a couple of hundred posters in this competition.

The Greenwich girls are standouts and it's clear why Andy said, while trying to manage Steven, "I don't worry about the girls." If they weren't at the fair, they could be lunching at the club. Olivia joked that her science fair look is that of a flight attendant—a blazer over a dress. She and a few others have chosen this attire, while some wear trim sweater cardigans over their dresses. Most wear short heels. With their hair blown straight and pushed behind their ears to reveal tasteful stud earrings, they're so put together, they sparkle.

Sophia's wearing a navy blue Club Monaco sweater dress with a small kite-tail cutout pattern running down the shoulders. She has perfectly matching suede pumps. In this ensemble, she looks considerably older. Connor's in khakis, a tie, and a blazer that likely fit him better last year, as now the arms are just a touch too short.

You stop when you see Steven, who's gently pacing in front of his poster. He's nearly unrecognizable in crisp khakis that fit him perfectly and are not high-waters, a spiffy shirt and tie and a blue blazer that's several shades lighter than navy. He's got on new shiny black shoes. He did it. Or someone, likely his mother, did. Regardless, he looks great. And his hair, which has always been cut into a soup bowl, is nice and fluffy today. It's a whole new Steven.

All the kids, Andy's and others, are jittery. Some practice by actually miming and monologuing their entire presentations, scrunching up their faces when they make an error and then starting over. Andy's kids fidget and seem not to know what to do with their arms and hands. They'll each get two rounds of judging by teams of two or three people at designated times. The judges, about two hundred volunteers from the community, view their posters and then listen to the kids' presentations and ask questions.

Andy walks around with his Nikon shooting photos. He keeps the mood extra light and doesn't engage in any pregame pep talks, other than to say, "Good luck, you'll be fine." The kids are all deep in their heads and don't seem to want anything in the way of a chat.

The judging teams have an itinerary of projects they're assigned to view. They approach the students and read the posters, trying to get a grasp of the project. Andy's kids are ready.

Connor developed a water filtration system using boron nitride nanotubes instead of the more commonly used carbon nanotubes and built a plastic membrane that used magnetization to increase efficiency and speed of the filtration process a thousandfold.

Connor's a self-assured, smooth public speaker. He walks the judges through each step of his work. He even built a 3-D animation on his laptop that demonstrates how the boron nanotubes work. One impressive aspect to this fair is the engagement of the judges. The fair organizers recruit scientists, teachers, and professors, all of whom take the kids' work very seriously.

A guy judging Olivia listens intently to her presentation and says, "Do you have a patent for this?" (She does for her Ebola diagnostic, but not for her Lyme one.)

One of Shobhi's judges asks whether her project considers proteins or protein fragments in the diagnostic process, a point that Andy warned her days before she should be prepared to answer.

Once they're gone, Shobhi says to Andy at about sixty miles per hour, "Oh, my god. This one guy asked me whether it uses proteins or protein fragments!" Thanks, Mr. B.

Steven's first round of judges is slated for ten thirty A.M. Andy has worried that judges either won't get Steven's project or that packing as a topic won't ignite much interest. But Steven's initial audience has a scientist who hangs on every word exiting Steven's mouth. He's firing questions at him.

"Have you thought about low-density packing?" the judge asks. He works in 3-D printing and thinks Steven's project could have some industrial applications. The judge has an awestruck look on his face, like he can't get enough of Steven, who naturally can answer any question possible on the algorithms related to packing. The team lingers for about thirty minutes. Later, in the cafeteria, the judge walks past Steven, gives him an attaboy pat on the shoulder, and says, "Great work, Steven."

Steven has a long two-hour wait before his next judging round. A visual spot-check on him prompts a double take. He's engaged in what looks to be a fairly intense conversation—with a girl. A few minutes later, Connor joins the conversation. He's completely oblivious that he needs to scram.

Andy can't help himself. When Steven has resumed his gentle pacing, Andy bounds over to him. No preamble.

"Did you get her number?" he asks.

"No," says Steven.

"Why not? You gotta get the number, man. Did she seem interested?"

"In the poster, maybe," says Steven. "Yeah, my poster."

"Are you happy? Did you like her?"

"I don't know," replies Steven. "I don't think so."

Andy gives him a jocular shoulder slap.

"You're not getting this at math team, buddy." Andy laughs and walks off.

Even if Steven isn't writing sonnets, it's a rather historic moment.

Around one thirty, everyone files into the auditorium for a keynote speech and awards. All ten of Andy's kids sit in one long row. Some of the other schools have brought so many students, they take up entire sections of the auditorium. Since Andy only brings kids with completed projects, as opposed to project proposals, his crew is considerably smaller.

From the roughly 225 entries, the judges award prizes from honorable mention up through first in four scientific categories— behavioral, physical, health and medical, and environmental. There is also an array of specialty awards given by various organizations, including one from the Office of Naval Research that Devyn Zaminski shares with four boys. For her project, she'd developed a sustainable way to clean water while simultaneously creating usable natural gas.

Finally, the main awards begin. The behavioral science winners are first and Greenwich isn't called. Everyone's facial muscles tense up.

Next up is environmental science.

"Third place goes to Devyn Zaminski, Greenwich." Devyn makes her second trip to the stage.

". . . And first place goes to . . . Agustina Stefani, Greenwich."

Andy's eyes pop open and his jaw drops.

"Oh, my god!" he mouths. He's stunned. Agustina practically leaps onto the stage to collect her prize for her project, which used a water solution to capture carbon dioxide from the atmosphere. This marks Agustina's very first win after sitting on the sidelines while so many of her peers claimed victories.

There's a marked difference between the demeanor of the Greenwich kids and that of everyone else. When the other schools' winners are announced, the kids go wild, shouting and whooping, some even jumping out of their seats. Andy's kids do a polite golf clap. It's unclear why they don't engage in any raw displays of emotion. Sitting there, they look like figure skaters who have come off the ice and are awaiting their scores, a look that is a mixture of poise, shock, and the brink of great joy or collapse. They look vulnerable and younger, shrunk back to their teen selves after putting on their grown-up presentations.

Next up is the toughest category for Andy's kids, health and medical. He's got so many prizeworthy kids and projects, each one more impressive than the next, this is the real nail-biter for him.

The announcer says, "There's a tie for honorable mention. Sophia Chow, Greenwich." Sophia, looking stoic, jumps up and walks to the stage to collect her medal.

"We have a tie for first place. Olivia Hallisey, Greenwich, and Shobhita Sundaram, Greenwich."

Andy nods knowingly and smiles.

In the physical sciences, "third place goes to Steven Ma, Greenwich." Steven's sitting with Connor and Manuel Carballo and they slap him on the back.

"And first place goes to Connor Li, Greenwich." (He ties for first with a student from Darien High School.)

Andy beams and runs up to the front with his Nikon to snap photos of his kids.

The fair folks then give out a slew of additional specialty awards and nearly all of Andy's kids get called. But the moment everyone's braced for—the announcement of the three kids who will be part of Connecticut's delegation to Intel ISEF—is yet to come. Since this is Andy's first time bringing his kids to today's fair, he's not sure how exactly this decision is made. His best guess is that they're going to pick three kids from the first-place winners—which means Agustina, Shobhi, Connor, and Olivia are all in the running.

Fair director Zia Mannan steps to the podium. Silence falls over the auditorium.

"This is a bit like the Oscars. I haven't looked at the winners, I don't know the results," he says to nervous laughter. "I'm just going to read the names."

"Agustina Stefani . . ."

Andy's eyes nearly fall out of his head. Agustina walks slightly more slowly to the stage this time, likely from shock. For her first-ever win on the science fair circuit, it's a big one.

"Shobhita Sundaram." Shobhi stands up, walks to the stage, and takes her place beside Agustina. Andy's clapping and shaking his head.

"And Connor Li." Connor bounces up from his seat, throws his arms up, and runs onto the stage mid fist-pump. Agustina and Shobhi are waiting for him and he throws his arms around Shobhi in a jubilant hug. Finally—some unabashed triumph. There's a loud, collective "Awwww" from the audience.

Greenwich has swept all three Intel spots—and it's their first time ever at this fair. When the kids descend from the stage, there's a lot of hugging and congratulating. Agustina starts to cry.

"I can't believe it. Oh, my god. I'm in shock. I'm so happy," she says, her voice cracking. She gives Andy a big hug. Her father approaches and he and Agustina start chatting in Spanish. He's the picture of paternal pride. Agustina has an interview for Penn today at four P.M., so she's got to keep an eye on the time. But she sticks around long enough to bask in her victory.

At a time when our political rhetoric is larded with talk of wall building and travel bans and deportation, it's an especially poignant moment to see three first-generation, publicly educated kids take the stage to claim the biggest prizes. Agustina's parents emigrated as adults from Argentina. Shobhi's parents are of Indian descent, having been raised in India and England. Connor's mother is from Hong Kong and his dad grew up in New York City's Chinatown. With their families hailing from distant points on the globe, these three are standing on the stage bolstered by their parents' aspirations and dreams to give them the best. They have lived on a somewhat peripheral Greenwich ring. They don't dine at country clubs or attend charity galas. But they chose America and then Greenwich for the amazing public education, for the resources and prospect that their kids might be able to do awesome things. They hoped there might be an Andy Bramante who would give their kids an unbelievable chance to grow and learn and discover a whole new part of themselves.

And today, it feels like they got it all.

36

Olivia

When the forty Regeneron finalists are named and Olivia is not among them, Andy's not very surprised. He observes that Olivia has won so many accolades for her Ebola diagnostic, the science fair world might feel she's gotten her due.

But what happens to her following the Regeneron announcement shows a sordid, ugly side to the übercompetitive science fair sphere that has given her such success.

On the heels of Olivia not making finalist, some kids take to online message boards to say cruel things about her and rejoice in the fact that she isn't advancing. The remarks are personal and mean. For some, it seems her star is too high, too bright. Again, this is part of the price of becoming a high-profile student celebrity. Most of the comments are deleted as quickly as they appear. But for Olivia, the damage is done. Despite her public persona of being immune to such craven behavior, she simply isn't. These words are wounding and she can't fathom why people are so jubilant about any of her perceived flaws or failures. The obvious interpretation is that this kind of anonymous sniping is fueled by envy and jealousy. But there lurks a current of contempt as well. Which is odd—feeling hateful toward a science fair superstar seems kind of pathetic.

Cyberbullying knows no boundaries. Even being the happy popular athletic whiz kid can't shield you from it. In fact, it's almost certainly because Olivia is the happy popular athletic whiz kid turned media darling that she's become a target for haters. It's just hard to imagine people pouncing on a nerdy, awkward incarnation of Olivia. Which, of course, brings us back to the female-specific dynamics— the ostensible impossibility that society will fully embrace and celebrate a strong, substantive, accomplished young woman.

So it's not surprising that despite her first-place finish at CT STEM (her tie with Shobhi in the health and medical category), Olivia decides to bow out of competing. She withdraws from CSEF, saying that it's time for others to have their moment. It's understandable and some might say a gracious move on her part. But it's also a total shame because it seems like she's exiting the fair world in part due to all the rancor.

Some of that rancor is in her own backyard. Following CT STEM, Andy spends forty minutes on the phone with a mom who is highly displeased that her child didn't do better at the fair. One of her questions, which she pointedly asks Andy, is, "Why was Olivia there?" She's partially scolding Andy for allowing Olivia to compete, insinuating that because he did so, others were robbed of a chance to win. He isn't having it. He reminds the mother that she didn't feel that way when one of her older children, a dominant force on the competitive science fair circuit, entered again and again.

As for the parents of the star competitors, they're no dummies. They know that as their kids' wins pile up, a target appears on their backs. Bonnie Novek, Ethan's mom, is self-conscious that people might think Ethan or their family are greedy if he continues to compete for high school science money while he's also vying for $7.5 million at the Carbon XPRIZE.

Andy says Olivia's hitting capacity with all the notoriety and that she's looking forward to college where "she can disappear." At this moment, the downsides to fame are outweighing the benefits.

The only place now that she truly feels she can disappear is the pool. Every day after school, she heads to Chelsea Piers in Stamford and practices with her team for two hours, swimming what seem like endless laps. In the pack of rubber swim caps and goggles, it's tough to pick her out in the crowd of lithe teens. But this is her nirvana.

"I need it to clear my head," she says. The monotony of the laps has a meditative effect, one that helps her momentarily fade and forget about Ebola, science fairs, college, etc.

As for college, she too is offered an early spot at Stanford, despite applying regular decision. She initially decides to keep this under wraps, for fear that people will think she's bragging. But a friend of hers wants to post a congratulatory message on Facebook—and Olivia lets her. In the end, she decides it is something to be proud of.

Like William, she doesn't jump to immediately take the offer, but she admits, "I want to go somewhere that wants me," rather than feeling like she's begging on the doorstep. Unlike William, she doesn't attach a disproportionate importance or meaning to getting into Harvard. However, she knows that if she goes to Stanford, with its big-time athletics, she won't be swimming competitively.

Her parents, Julia and Bill, steadfastly refuse to weigh in on the matter. (Bill went to Stanford as an undergrad.) "It has to be her decision," says Julia. She's a firm believer that Olivia needs to be happy wherever she goes and that that's not something she and Bill can mediate.

"When I've had troubles or challenges, I've learned that the universe works in different ways," says Julia. "The only thing I worry about is health. Everything else can be taken care of."

In the end, Olivia has several great college options. But the one that feels the best is Stanford. She's known for a while that she'd like to pursue medicine. She certainly has the stamina for medical school and the punishing hours of residency. And she has the inspiration and example of her grandfather, whom she sometimes googles when she finds herself deeply missing him. Of course, much can change in college and she may find some other path.

Her scientific accomplishments have sometimes come at a price—it has seemed as if Olivia herself was often put under a microscope of public opinion. Which is why she's looking forward to stepping out of the spotlight, moving across the country, and starting fresh.

37

Danny

"Fume hood? Fume hood?" Henry Dowling says to Danny, who's working with toxic chemicals out in the open.

"What?" says Danny, not looking up.

"We have two nice fume hoods here. Don't kill us all," replies Henry. He has an amusing way of minding, chaperoning, and needling Danny, always managing to insert himself immediately prior to something potentially catastrophic.

Danny's goggled and gloved and has found his battery. It was in a case on top of the glove box. How he didn't find it during the great search remains one of the universe's unknowns. "It's still a good question of where it went," Danny says. "I checked there not once, but seven or eight times."

The problem is that when Danny located the battery he'd built for testing, the casing was separated and the petri dish was filled with white powder.

"I must not have crimped it hard enough and I think some air got in there, oxidized the lithium, which is why I had to do it in a glove box to begin with," explains Danny. "And it seems to have exploded."

Christo Popham, a fellow senior who built his own battery the year before, is watching Danny work. Christo has a look of deep concern on his face as he watches Danny delicately placing bits of silicon wafer into a fresh battery casing.

"You're using fragments?" says Christo.

"There's no way I can get—" Danny starts to say.

"I took a chisel and cracked it in specific places," says Christo.

"It won't work, too brittle."

"If there's gaps in between the pieces, not good," says Christo, shaking his head.

"What happens then?" asks Danny.

"If there's spaces in between, lithium would be able to bypass the silicon altogether," says Christo. Translation: if the silicon fragments are not packed tightly enough into the casing, it's going to make for a mess and the battery won't work.

"If this doesn't work," says Danny, "I'm going to have to skip CSEF and just go to Norwalk."

Danny's trying to manage some other crises as well.

"After I got into college, it's been the most stressful semester of my life. I am exceptionally good at fucking things up. Just about anything," he says. By way of example, he points out that he almost lost a class officer election freshman year—even though he was running unopposed. Something having to do with an email snafu. But the point is taken—who nearly loses an unopposed contest?

It devolved into an argument with a teacher. When he was allowed to run and make a speech, his opening line was, "Welcome to my Soviet-style election."

"And she wanted me to get rid of that line because it was offensive to Soviets, to which I argued the Soviets don't even exist anymore."

He went on to say in his speech, "I'm the best choice, the only choice."

Three years later, he calls the whole debacle a "formative experience in my fight for free speech."

His current albatross is a conflict about a class. Harvard insists he must take an honors English class this semester—and isn't budging. But Danny's schedule somehow isn't conducive to this and he's tried switching out of a few classes to make it all work. He keeps shuffling around options, all of which are precarious and require concessions on a few levels.

"I feel like I'm just going to do something and just fuck everything up. Somehow, some way, the world is going to find a way to take all of this away," he says. "Deep down, I really just think I'm going to screw it all up and have no college."

While managing his perceived sense of doom, Danny's also trying to tick some of life's basics off his list. His road test is imminent. There's somehow a lot riding on this. He can decline Latin verbs, get into Harvard, and do differential equations with his eyes closed. But if he doesn't pass his road test, in his mind he's not yet made it to manhood.

In advance of the test, he practices driving around Greenwich, staying within the speed limit and then doing all the various turning and parking maneuvers.

The Monday following the test, Danny offers the following report:

"I failed the judgment part of the driving. Judgment is a *basic* driving skill, not a *critical* driving skill. My driving was fine; I failed the judgment while parking. I figured I was on the line and I knew how to correct it and I didn't know at what point he'd allow corrections. I put my foot hard on the brake and opened the door. I didn't put the vehicle in park before I opened the door. It wasn't a danger, but it wasn't something I was supposed to do.

"I failed acceleration because I'm bad at staying within the speed limit."

The guy assessing Danny's performance called his driving "jerky." As in not smooth. It turns out acceleration falls into the category of basic driving skills and "you can fail up to two basic driving skills, but none of the critical ones, like parking, or stopping at a red light."

Despite his infractions, Danny earned his license. "I passed it, but barely. I don't get my license for real until tomorrow," he says, adding that he still doesn't plan to drive himself to school. "I don't trust myself to drive in the morning because I'm a mess."

Danny says all this with a wooden spoon in his hand. Assassin, the senior elimination game, is now full on. No more beta, during which he finished in sixth place. This is the real thing. He's out for blood. Danny's still slightly bitter about what he says was poor intel from Rahul during the beta phase.

"You gave me incorrect information," Danny says to him. "You told me Mark was after me. Thank you for the info, Rahul. Have another slice of pizza. I thought you were my friend."

Even though he worked out his honors English class situation, Danny now finds himself flailing a bit in some of his other classes.

Diagnosis: acute senioritis.

"I grew complacent at the beginning of the quarter. I'm hanging on by a thread in a few classes. My grades are always absolutely horrible until the very end of the quarter," he says.

He's decided to put off his battery until after CSEF, seeing as he wasn't going to make it anyway. He says he'll focus on the Norwalk Science Fair, the starter fair he went to as a sophomore. The goal, in his mind, seems to have morphed from which fair he attends to just getting to a fair—to complete a project.

Today, the exemption during Assassin is dancing. That is, if you dance, you cannot be killed off. So Danny leaves the lab, wooden spoon in his left hand. With his right, he does a poor man's version of the disco move from the 1970s, where he raises his right index finger in the air and crisscrosses it down to his left hip. In a rarity, when he walks down the hall a few minutes after the bell has rung, he has the corridor to himself. But he's leaving nothing to chance. Plodding along to his next class, bathed in midday sunshine streaming in through the windows, he protectively does his disco move the entire way.

It's April 6, an "E" day in the school schedule, which means pizza day, courtesy of Danny. But he's at a math meet, so his pal, Henry Dowling, has stepped up and offered to be the pinch hitter pizza man. Henry quickly learns that getting contraband pizza into the school is no joke.

"Danny has this intricate network in place to pull off pizza day. It's not easy," says Harvard-bound Henry. "He has the guy at the pizza parlor who has it ready for pickup at exactly 11:37 A.M. He has someone pick it up and then somehow gets it past the security guard. His pickup person was not available to me."

So when Henry finally walks in with the pizza, he's quite stressed by the whole thing. Not to mention an impatient and eager crowd awaits. Of course, there's nothing in the way of gratitude. Everyone's been spoiled by Danny.

The following day, Danny gloats, saying to Henry, "It's not so easy, is it?"

Danny's headed to Copenhagen for spring break this year, which he's looking forward to.

As for that battery, Danny says he's going to get to finishing it "soon." He confirms the date of the Norwalk Science Fair, April 28.

Uh oh.

That date sounds familiar. Danny realizes that he has a math meet that same day. He's one of the team's co-captains.

"Yin can't go," he says. William is also one of the captains of the team, and he's also one of its strongest members. If Danny also doesn't make it, well . . .

"I may email the Norwalk folks and ask if I can present early," he says.

This all feels a bit premature, seeing as how he still has not (re)constructed his battery.

Andy thinks it's a foregone conclusion that Danny's battery will not see the light of day. He thinks Danny walked the plank somewhere back in the fall. He's disappointed, for sure, though not surprised. He feels like a bit of a schmuck for getting charmed into thinking this would be the year that Danny got something done.

"It is what it is," he says with an air of resignation so thick you could write your name in it.

38

William

William's early acceptance to Stanford, despite his having applied regular decision, is a move by the university to nab exceptional students. Which is why it's not surprising that Olivia also received an early spot. Stanford offers William an all-expense-paid trip out to Palo Alto, California, to visit the campus with other early accepted students.

"I was really surprised," he says. The Harvard deferral so knocked him off his game, it seems to have impaired his ability to think anything good on the college front would ever come his way. He's intrigued and keeping an open mind, but he's not jumping to say yes. In fact, he's completely noncommittal. A couple of weeks later, when he gets a similar early acceptance to Yale, to its elite science program, he can barely muster a smile.

What gives?

Well, he's still holding out hope, despite being the one who initially stated that his deferral was a death knell, that Harvard will take him. What's unclear is what's driving this. He's the first to say he wasn't wowed by Harvard when he visited. It sometimes feels like he simply can't take the snub—that he's at war with the inherent unfairness of it all. What never comes out of his mouth is a declaration of awe for the school—or why it's the best place for him.

William's operating with KGB-like secrecy about his Stanford acceptance, wanting no one to know. He's developed conspiracy theories that run from outright sabotage to accidental disclosure. He's afraid that someone might call up Harvard and tell them he's been accepted to Stanford, prompting Harvard to reject him. He outlines scenarios in which a casual conversation overheard among the wrong people could lead to word getting back to Harvard. This all assumes that Harvard is keeping a constant pulse on one William Yin, that there's a WY antenna and the university is frequently turning its dial to try to get updates.

Andy finds this maddening. For one, the fact that acceptances to two major, prestigious schools yield no enthusiasm from William strikes him as ludicrous. Most kids would be euphoric to get a spot at either, let alone both. He firmly believes William needs to let go of his Harvard obsession and turn his attention elsewhere.

This whole thing cuts close to the bone. Andy's daughter had her heart set on Tufts and didn't get in. She's now a student at the University of Delaware and says, "I can't imagine going to school anyplace else." She doesn't just like Delaware, she loves it.

Things got a bit ugly between Andy and William a few weeks ago, in the run-up to the CT STEM Fair. Given the complexity of William's Alzheimer's diagnostic project, plus his trademark procrastination, it became clear that despite applying and being accepted to the fair, he wasn't going to be able to finish his project.

But because of his myopic determination to show Harvard something that might turn the tide, he called up Andy the night before the fair and tried to make a case that he could present his concept at CT STEM, in lieu of finished work. Plenty of kids from other schools do this.

William's mere suggestion of such a thing infuriated Andy—who shut down the proposition immediately and unequivocally.

"William, don't even think about it. First of all, that's not what we do," Andy told him. "And second of all, that would be totally unfair to all the kids in the class who have busted their asses to get to this fair. Forget it, it's not happening."

What Andy didn't say to William is that he also barred him because "he's such a great speaker, he can really turn it on, it's a novel project and he could very easily walk in and win it without having done the work."

Andy says he feels sick at the idea on two fronts: that William would float the idea of presenting a half-cooked project at a fair for no other reason than this college crap, and that if Andy were to somehow allow him to compete, he'd possibly win over a much more deserving kid. He's also insulted by the ask. His program is distinguished for many reasons, one of them being that the kids never turn up with a concept or partially done research. And there's no way in hell that's about to change for William Yin.

Once Andy shuts down William's attempt to go to CT STEM, William knows his last shot is CSEF, the legacy state fair. He's got about five weeks to complete the project. Since the only thing he currently has to show for himself is the formation of an impenetrable blob, he's got some serious work to do.

But as has been his habit, without the flame of last-minute urgency burning beneath his rear end, William somehow waits until the week before the fair to tackle his project.

Remember, he has to construct two microscopic boats, Class A and Class B, that would theoretically travel to the brain. If the boats encounter the oligomers associated with Alzheimer's disease, the Class B boat will destabilize the environment and split open the Class A boat, which will send chemicals into the system that will prompt a color change on a urine test to indicate the presence of Alzheimer's. He also needs to show that curcumin—the natural substance believed to cross the blood-brain barrier—will actually bind to the oligomers. And he needs to build and test a color change strip.

None of these steps are easy or straightforward. But they now have William's full attention and the added adrenaline boost thanks to his back being up against the wall.

Since his first attempt to build the polymeric shell—the outside of the boat—yielded the blob, William decides to contact the researcher at the University of Edinburgh who wrote the paper involving polymeric shells. He wants some advice, some guidance on what he might be doing wrong and why it is that all he has to show for himself is a slimy blob. But what are the odds that a scientist on another continent will spend time answering some high school kid's questions? He probably won't. William knows this and he comes up with a shrewd solution.

Thanks to his summer program at MIT, he has a functioning MIT email address. So while he never states he's an MIT student, he begins a correspondence with the researcher, who actually gives William some invaluable insights.

Numero uno is that William's polymeric shell for the boat will never work for his experiment. His hunch that he could build one from the paper he'd read is wrong. The researcher shares a relatively straightforward (if you're into this sort of thing) methodology for the polymeric shell William needs.

"His process is a lot simpler," says William. But it's still very involved. In fact, building a polymeric shell, it turns out, is a complex organic chemistry experiment that's meant to be done in a specialized lab with isolation chambers. Since those conditions aren't available to William, his makeshift fix is to work in a giant plastic bag, an industrial-sized clear sack with built-in arm and glove sleeves. The bag covers the entire surface of a lab table and has quite the *CSI: Greenwich* look about it.

The entire process must happen in a nitrogen gas environment. Lots of projects require nitrogen, and Andy has two tanks in the lab that he regularly gets refilled. But William's project also requires that he use elemental potassium, a highly volatile compound.

"Pure elemental potassium mixed with water could blow a plume the height of the school," William casually mentions one day. Blowing up the school would certainly put a dent in his project, so he proceeds with particular caution.

Andy's far from thrilled about this setup, but he trusts William. There are some kids to whom he won't hand a book of matches unless he's standing beside them wielding a fire extinguisher, but he knows that William fully understands the ramifications of any carelessness.

Much to everyone's relief, William builds what he hopes is the polymer sans explosion. (There is no blob this time.)

But the telltale sign of whether the polymeric shell has been formed is a color change that should happen after the concoction of chemicals is combined and stirred. The whole thing should go from a slightly milky yellowish fluid to green to black. And it should happen quickly.

William's new problem is that once he's combined everything in the glove bag, his liquid yield is very small. He has just three milliliters of his concoction. When the kids need continuous mixing, they often use an automated stirrer in the lab. A metal stir bar (the size of a small twig) is dropped into a beaker holding liquid. The beaker is placed on an electric stirrer that uses magnetic force to whir the bar around.

William pours his mixture into a beaker and drops the stir bar in, but there's so little fluid that it all kind of gathers up around the bar and doesn't stir properly.

And this is where Andy comes in. He immediately sees the glitch. One of the problems is the stir bar isn't the full diameter of the beaker. He suggests William pour his liquid into a much smaller vessel that's about one centimeter in diameter. But they don't have a stir bar that small.

So Andy suggests they MacGyver one themselves. They cut a 1-centimeter piece of a paper clip and place it into thin glass tubing.

They seal the ends by placing them over a Bunsen burner, rounding off the sharp edges, which prevents any of the precious liquid from seeping into the stir bar encasement.

William places the jerry-rigged bar into the beaker. It starts stirring and within minutes . . . the liquid changes from its milky yellow to green to black.

It works. In all senses. When William witnesses the color change, he can't believe what he's seeing.

He adds the gold nanoparticles, and the polymeric shell naturally forms around them. William verifies this by taking measurements of the diameter. As the diameter of the boat increases, it shows that its walls have expanded to hold the contents.

He then needs to see if tetrazine, when introduced into the mix, will break apart the shell. The whole mechanism of his project depends on whether the presence of oligomers in the brain will trigger the tetrazine to release the chemicals that spark the color change on a urine test. He places tetrazine into his mixture.

"I checked their diameters and the one with the tetrazine had dramatically smaller diameter and it just contained gold nanoparticles," says William. The one without tetrazine underwent a negligible change in diameter.

He next needs to show that curcumin, the substance that appears to be able to break through the blood-brain barrier and bind to an oligomer, does, in fact, do this. He purchases oligomers and then introduces curcumin. He takes chemical measurements of curcumin by itself, the oligomer by itself, and then the two combined. Indeed, they bind together to create a larger mass.

He finishes the curcumin test thirty-six hours before CSEF. He still has to construct his paper-based test to confirm that a color change will take place if the oligomers are present in the brain. And he hasn't even begun to write up his poster—the showpiece of a science fair project.

At this point, all of Andy's other CSEF-bound kids have their posters. These are the slick, professionally printed, four-foot-high, three-panel posters mounted on foam core that have come to be associated with Greenwich and Andy's class. (Senior Margaret Cirino calls the posters "iconic.") Since William is so late, he's on his own to figure out the printing.

"William's first year, I almost didn't take him," says Andy. "You see him, you see what he does. Everything is last-minute, last-minute. And he was the last one I was considering next to a junior. He was a sophomore. I said, look, I've got to give this junior a chance. But [William] really turned it up, he was here every day. The kid who was a junior just remotely cared, he didn't really care. And so I took William and he wound up going to Intel."

Another way to look at it is that in going to CSEF and then Intel, where as a sophomore he won a second-place award, William was handsomely rewarded for his bad habits, which he repeated in much more extreme fashion his junior year. That's when he cut all of his classes for two weeks to focus exclusively on his plaque detection sticker. He was doing the backstroke in a swamp of misery, reduced to tears at one point, and delirious. But again he prevailed and went to CSEF and then Intel, where again he received a second-place award.

And so, here he is in his senior year, taller, possibly wiser, but pile-driving himself into a kind of mania that he both deplores and relishes. It's become his thing.

For the very final experiment, he has to construct his test strip, which is the piece that would ultimately give the diagnosis of Alzheimer's.

"I read a lot of papers about test strips," he says. Compared to the polymeric vessel, this step is like making a paper snowflake with careful scissor work. He dips paper into sugar water and some chemicals and then bakes the paper in the lab oven.

"Like a hard cookie test strip," he says. If it works, once he dips it into the telltale solution, the strip should change color.

Standing in the lab, he gently lowers the strip into a beaker and then watches as it turns from white to dark blue.

William Yin has done it again.

39

Romano

It's Tuesday, February 7, and Andy and the troupe he brought to CT STEM are still decompressing. With the season's first competition behind him, Andy's giving a sobering talk about the onslaught of deadlines and fairs that lie ahead.

"The Davidson application is due tomorrow," he says. "If you haven't started it, I'd take a pass. The twenty-fourth is ISWEEEP. Basically if you're near finished, they tend to like us a lot. Every year, we get about five kids going. It's not a problem for me, but Shobhi just mentioned to me that it's the week prior to ISEF. So there are issues there with AP exams. You guys that have been through this, you know this."

He's posted his infamous color-coded status chart and he tells the class, "If you're in dark green, you're pretty much done. I mean, there are some of you that are not done, but more or less done. Light green in my mind means I'd like you to go, but you've got to show me some data and I'm supposed to decide by Friday, so there's no pressure here. And then if you're in yellow, man, you'd better do handstands between now and Friday because ultimately I can only send eighteen.

"Know that, and I'm not threatening anyone, if you're in light green, and you don't get anything done, you can easily fall off the cliff. It's time to shit or get off the pot."

He also uses his monologue to caution kids about pestering him for expensive materials at this stage. "I have very little money for those of you who keep telling me you need, you need. Start a GoFundMe. Get a cup and go sit on a highway. I don't care.

"So anyway, that's the deal. I'm going to eat my salad now and relax."

The last syllable of his last word hasn't fully left his mouth and Romano says, "Mr. B. . . . ?"

"I didn't even hit the fork and that corporal punishment already started," says Andy, referring to the constant staccato of his own name.

"I know, I know," says Romano. "You can even just keep eating. Do you mind if I explain?"

Perhaps, but first Andy wants to take a moment to acknowledge the delectable tomato sauce Romano brought him, courtesy of Nanni, of course.

"It was outstanding," says Andy, who, of course, knows his Italian sauces.

Romano's pleased and takes the compliment, but he's got a look of seriousness and urgency. He launches into his own monologue.

"The biggest problem that I have is I'm going to have this gelatin in a tube. Then you're going to squirt it onto your arm and it's going from a medium temperature to a higher temperature—the surface of the skin—and pretty much, long story short, there's no way for it to be liquid in a jar and then solid on your skin. I've been trying to figure it out for so long and the DOPA, I've been doing a little bit to see how it affects curing time and it does affect curing time . . ."

"Not that quick," says Andy.

". . . And if it's going to be solid on your skin, it's going to be solid in a tube. I don't know. I'm at an impasse. I don't know. I just can't figure out how to . . ."

Andy's plunging his fork into his salad, mowing through it with determination.

"There's a lot of wisdom in this salad. Why not then back up and make a bandage where it's already hardened? Why does it have to come out of a tube?" he asks.

"What do you mean?" says Romano.

"You know how a Band-Aid's got the little pad on it? Why not have your pad be your gel? Make a new kind of Band-Aid. It does exist, by the way, a liquid bandage. Someone mentioned it to me totally unrelated to you," says Andy.

Romano has already researched this. "It's like moleskin, you can like spray it on. The only problem with that is . . . you know, like drug delivery? There's nothing with that because it's just such a hard polymer."

"It just becomes a hard polymer?" asks Andy.

"It literally just becomes plastic on your skin. So if I could make one that's organic and that provides benefits to the wound . . . that would be big, but I can't friggin' figure it out," says Romano.

Andy appears to think this through. "So what I'm thinking is why does it have to go from liquid to solid? Why not just embed the drug into the polymer, make thin sheets of it, double-side-tape it to a Band-Aid? I mean, it's not as sexy, but . . ."

"I don't know, I was going to figure out the drug to help the wound later on. My big thing was squirt it on and let it sit. So I could do that, but I haven't started as much research in the drug that would help the wound," says Romano.

"You in a rush?" says Andy.

"Well, honestly, the science fair, do you think that's even . . . ?" asks Romano.

Here, Andy's totally frank.

"You're not going to make it," he tells Romano.

"That's fair," says Romano.

"You're not even going to be here next week. But that doesn't mean you stop. You've got the Talent Search next year, there are other fairs," Andy offers.

"Can I keep researching after the [state] science fair?" asks Romano.

"You can do whatever you want."

"I can keep going?"

"What does it mean? I kick you out?"

"I don't know. I have no idea. Because I would like to do that."

Andy then reminds Romano of all the possibilities down the road, both this year and next.

Romano says, "The science fairs aside, I'm actually having . . ."

"You having fun?" asks Andy.

"I'm having a lot of fun," says Romano.

"It's fun, but I want you to really . . . I would love for you to get past this notion that you're spinning your wheels—and you are a little bit," says Andy. "Not for lack of effort."

"It's been months," says Romano.

"So you're of the mindset now that 'I promised I was going to do this.' Promised who? You know, no one? Yeah, your application for the state fair, but who cares about that? If you need to change gears, you change gears, man, that's what it's about," says Andy, all easy breezy.

Romano floats a new direction by Andy, one that involves different chemicals.

"But I'm thinking again of the science fairs later on. Is this something that could turn into something or is it just like . . . ?" says Romano.

"Are you interested in it?" asks Andy.

"I am, yeah, I like it a lot."

"So who gives a shit? If I say no and I say let's do how to clean a tomato jar . . . well, you're not interested, so what the hell's the point? So yeah, if this is . . . if you're digging it, we'll make something work. It's a moving target right now. The only reason you're feeling pressure frankly is because you think you've got to hit this target for some fair director."

"It's the time, it's the time," says Romano.

"If we let go of that, then you can do whatever you want," says Andy, still munching on his salad.

"Yeah, that's a good point," Romano finally concedes.

"I feel like I should have a couch," says Andy.

"A couch? For you or for me?" asks Romano.

"I do this with you, I do it with parents, I do it with everybody. Except my wife. She doesn't want to hear anything I have to say."

"It's really nice because everything nowadays has, you know, pressure, time. It would be nice to just do that," says Romano.

"I don't want you just putzing around," Andy says.

"Yeah, of course. There's a line, I understand that."

Verna, who's nursing a foot injury from running, has been patiently waiting her turn.

"Verna! So nice to see you. Gimpy Verna, how's your foot? What do you need?"

"Can you help me get a magnet off the wall?"

Romano briefly looks into using collagen as a base for his liquid bandage, but he decides it's best for him not to meander solo too deep into the forest, as he did these past few months.

"I'm just going to wait until after Mr. B.'s done with CSEF. He's got to focus on the kids going to that fair, which I totally understand, so I'm just going to take a break for now," says Romano. It's fully apparent that Andy is spread thin to the point of being two-dimensional, as if someone's pressed him into a life-size poster cutout of himself. Romano is accepting of the situation, if a bit forlorn. "Mr. B. took a chance on me and I failed and I'm just a burden to the class at this point. I don't have anything to offer. Compared to the other students, I'm not like a William Yin, taking all these APs. I'm a little insecure about my science stuff; I feel like I always have to prove myself.

"Out of all the kids in the class, I roll with a different crowd, I dress different, I'm, like, involved with a bunch of girls. I'm on rugby. I'm not taking twenty AP classes," says Romano. He's not wrong. He's definitely the only kid who says things like "I roll with a different crowd."

He's still thinking about his pitch to get into the class in the first place. "And my proposal was probably the worst out of anyone. If someone looked objectively, they'd say, that kid's going to fail and I failed."

Even though it can't totally compensate for his doldrums, Romano is looking forward to his winter break. He and Kim and one of his best friends are going to California to look at colleges.

As for his lady luck, he's been inching closer to Alessandra, the Italian senior, "the most beautiful girl in the school."

He ticks off all of their encounters, which include walking up a staircase together, sitting together in the media center. And then there's the big one.

She drives him home one day.

The vibe, to him, is undeniable. What senior girl drives a junior guy home? And the conversation just flows.

But still. He's got to play this one cool.

"You have to let things marinate," he says, with faux bravado, as if he regularly doles out advice on such matters. He doesn't want to be that eager puppy nipping at the beautiful girl's heels. These things take time. Rushing in is a sign of weakness. And there's one other minor wrinkle.

"I think she might have a boyfriend," he says one day, with a perplexed and pained look on his face.

Hmmm, this would seem like a detail worth uncovering, even on the down low. The thing is, he sort of wants to know and sort of doesn't. Okay, he truly doesn't want to know.

He tells Andy about the intrigue and cracks, "Just because there's a goalie doesn't mean you can't score." Andy seems to agree that he's better off not knowing.

The fact at hand is that she's made no mention of a boyfriend and she hasn't tried to throw up a wire fence between herself and Romano by doing the casual but deliberate boyfriend reference thing. He can't deny that it's on his mind and yet . . . their chemistry is soooo good, why pose the question and risk self-inflicting such a mortal wound? Better to ride the high of their vibe and just remain blissfully unaware.

But he's about to be brought down to earth in spectacular fashion.

When Kim, Romano, and his pal, Henry, head west, they hit the Bay Area first. The trip starts off with what Romano thinks is a promising omen. "Once we landed, I got out and there was an Audi and my mom was, like, that's our car," he says. "We usually get Traveling mom.com minivans."

The trip becomes a montage of cool, fun, impromptu California vacation moments. Romano gets a five-dollar haircut in San Francisco's Chinatown. They go to Fisherman's Wharf; they ride a cable car. They do a pedicab tour. They also visit UC Berkeley and Stanford, bumping into several GHS science research students at both.

In Los Angeles, he and Henry spot Jack Black. Henry's too shy to say anything. Seizing the moment, Romano walks past the actor and says, "Big fan." Jack Black tosses him a thank-you, a moment that shall forever remain frozen in time. "It's still amazing, beautiful. I melted. I love *Kung Fu Panda*. It's one of my favorite movies of all time," Romano says, oozing at the mere memory.

While in Tinseltown, they visit UCLA. Romano and Henry know exactly one student there, a GHS graduate. They used to work out at the same gym during football season. They aren't tight, but "I could wave to him in the hallways," says Romano.

Henry actually reached out to the guy, whom we'll call Joe College, ahead of the trip to ask if he could meet up with them. Romano and Henry are grateful Joe College is willing to take the time to show them around.

It's always entertaining to hang out with someone familiar in an entirely different setting. Joe College is super nice and gives the boys the deluxe UCLA tour. They chill for a bit in his dorm room. As they're wrapping up, they engage in some chitchat.

Joe College casually mentions that his girlfriend from back home has been out to visit. His GHS girlfriend, Alessandra. In this moment, asteroids collide, sending debris plummeting down to earth. You may have gotten pelted. Romano tries to hold it together, even though he wants to drive a steak knife into his heart and skull.

"She visited you?" Romano says. "Here? In California?"

"Yeah, she's been out here a few times," says Joe.

"Here? At UCLA?" Romano confirms. "That's cool. Wow."

There's a monologue running in his head and he has to fight the urge to keep talking. *Let me get this straight. Alessandra visited you? She flew across America to come here to visit you because she's your girlfriend? YOUR girlfriend. She's visited you multiple times. Alessandra, right?*

What are the fucking chances? That the ONE girl he's super into, let it be said again, THE MOST BEAUTIFUL GIRL IN THE SCHOOL, is dating the ONE guy Romano knows at UCLA . . . and he had to fly three thousand miles across the United States to find out this little humiliating fact.

It seems he's been played. He can still sort of reach through the abyss and meekly take hold of a silver lining, tarnished as it may be. He never outwardly tipped his hand. He didn't make any overtures that revealed his infatuation. From a public standpoint, his pride is fully intact.

Back in the lab, though, he's still whirling from the news, still in total disbelief.

He says, "I mean, what the hell?"

Luckily, since Romano's modus operandi since splitting with Caitlin has been to keep his options open, he's got another prospect. She's not the most beautiful girl in the school, but he's intrigued. A girl we'll call Mandy is the sophomore sister of the stage manager for

Noises Off, the play in which Romano is playing one of the leads. She sometimes hangs around rehearsals waiting for her sister.

Romano starts talking to Mandy and thinks the logical and easy way for her to maybe swoon is to see his epic performance. She might get a little starstruck, become a fan. Mandy's sister, the stage manager, tells Romano that Mandy hates theater.

Still, he embarks on a campaign to try to woo Mandy to the show. From there, he figures, things can only gain momentum.

Somehow, from the outset, Romano is off his game. For one, he develops a near tic—every time he sees Mandy in the hall, he immediately raises the issue.

"Hey, do you want to come see my play?" he says. "Play?" Since the changing of classes in a school so large is like a stampede, you've got to act quickly if you need to communicate something.

"The amount of times I asked must have exceeded double digits. It got to the point where I'd pass her in the hall and just blurt out, 'Play?'" he recalls. "I started creeping myself out."

He hasn't lost sight of the show itself and the need for his performance to shine. On the Saturday evening of February 25, the entire Orlando clan attends—his parents, Nanni, Rome's siblings. They sit in the center section near the front of the tiny theater, eager to see Romano's stage debut.

Noises Off, a farce, is a British play within a play. So in the role of Freddie, a sweet, dim-witted actor, Romano also plays Philip Brent, whom Freddie plays. Romano navigates an English accent better than most of the cast, but where he really shines is with all of the sight gags and physical humor. He's not afraid to drop his trousers to reveal gaudy silk boxers and rainbow socks. It's a robust and uninhibited performance.

Afterward, the *famiglia* gathers in the lobby, awaiting their star. Hugs and kisses galore, but none more plentiful than Nanni's. He has to bend down to hug her and she pinches his cheeks and then unleashes her arsenal of kisses. He giggles. Play or no play, Nanni is his forever fan.

Mandy never does come to the play, but Romano isn't giving up. Maybe it's because his ego is still stinging from the Alessandra flap and he needs to feel like he's still got it.

He decides to invite her to another student theater show called *Magic Circle*. Despite Mandy's sister having told Romano that Mandy really can't stand theater, this key piece of information somehow whizzes past his cognition, as if he were being told something on the tarmac while a jet was taking off.

So one day, he spots Mandy and a bunch of girls at a table in the middle of the student center during lunch, raising money for autism awareness. She's seated at the table. For once, they're not seeing each other in passing. This is his chance to lock in the date for *Magic Circle*. And, he's not going to lie, being that she's surrounded by her sophomore friends, he thinks he can cause a little stir. Cool older guy publicly asks younger girl out on a date.

He struts over to the table, all of his brrrn-chicka-brrrn-brrrn swagger on display.

"Hey, Mandy. I want to go to this *Magic Circle* thing this weekend and none of my friends want to go. Do you want to go?"

"When is it?" she asks, all logistics.

"There are two shows," Romano says. "Friday and Saturday."

"I don't think I can go," she replies.

Thud. She ends it right there. She doesn't fumble to fill the awkward silence. She blinks.

Romano's in a mental free fall. Wait, he thinks. She can't go Friday *or* Saturday? She's busy both days. Who's busy both days? No one is that busy. Maybe she's not busy . . . maybe. Oh.

"Okay," he says, trying to untangle himself from the parachute of humiliation that's piling up around him.

He starts to walk away and one of the girls gamely says, "Do you want to support autism awareness?"

"No, no, I don't," Romano sneers, full of indignation.

Later he indulges in some postgame analysis. "I was, like, I think for the first time, I was officially rejected," he says, struggling to fully comprehend this milestone. "Did that just happen? I have to think straight."

He calls his sister, Sophia, and recounts the whole sordid tale. "I got the big R," he tells her. Sophia laughs.

A couple of weeks later, he sees the humor in all of it and says, "When that girl asked me to support autism awareness, in my head, I was like, 'No, I'm not supporting autism awareness. First, I'm going to go to the corner and eat my sandwich. Then I'm going to go cry. Because this is the WORST FUCKING THING that's ever happened to me. No, I'm not giving you a fucking dollar!'"

40

Andy

With a possible blizzard looming the following Tuesday, the CSEF folks have moved the poster drop-off up a day to Monday. This means the kids are losing a crucial day. What it mostly means is that William has even fewer hours to finish his project, which has not quite progressed to a stage that merits the word "project."

Andy told the kids going to the Junior Science and Humanities Symposium (JSHS) that he'd be available on Friday if they wanted to Skype with him and do a dry run of their oral presentations. He liked the idea of holing up in his home and getting a day off from the lab. This year, William, Connor, and Shobhi have been chosen to deliver oral presentations.

Shobhi was the only one who contacted him on Friday. Connor didn't bother, in keeping with his stance that repeated practice makes him more nervous. William was radio silent. Andy texted him around five P.M. to check in. When William replied around ten thirty, Andy was already in bed.

The Connecticut JSHS competition format is notably different from the other fairs. As part of the application process, kids submit three-minute videos explaining their projects. Joy Erickson, who has been running Connecticut's JSHS for more than twenty years, says the point of the application video is for the judges to see the essentials quickly. "How good are you at talking? How good are you at answering the questions we actually ask? Kids will go on and on about their project. We ask four very specific questions." She notes that it's all about the content and not sleek video production. "We don't penalize kids who had bad equipment."

Those who advance are selected as poster presenters or the more prestigious oral presenters. Poster presenters give a more informal talk to judges who circulate around the room, much as at other science fairs. (A third group is also chosen to display their projects in an exhibition room. They're not vying for the larger prizes, though they are considered for the People's Choice Award.)

The oral presenters give twelve-minute talks in an auditorium in front of judges, their fellow presenters, teachers, parents, etc. Among other things, it offers the kids a rare opportunity to size up the competition. The kids prepare PowerPoint presentations and use laser pointers to highlight key ideas and images. They say the biggest stressor is the time factor. At twelve minutes on the dot, the judges cut them off, no matter where in the presentation they might be. Leaving nothing to chance, Shobhi timed her talk at home, and during her initial run, she clocked in at sixteen minutes. There's a five-minute question period from the judges and then the floor is open for anyone to jump in.

As with the other competitions, JSHS awards a bunch of specialty prizes, which are nice because it means many kids can go home with something for their work. Five top winners are named; oral presenters placing first through fourth and the top poster presenter are sent to compete at the nationals, which this year are in San Diego. Andy's had a handful of kids make it to nationals, but he says that level is excruciatingly tough. Only Ethan has ever placed—third—and he did that as a sophomore.

Rahul Subramaniam has been selected to present his poster for his Zika diagnostic. He and his father, Prem, are here. Another of Andy's students, Luca Barcelo, was also selected to give a poster presentation, but he's doing a semester at a specialized school in Washington, D.C., and couldn't make it up to Connecticut this weekend.

The fifty-fourth annual JSHS competition is being held at UConn Health, a teaching hospital. At noon, the students are congregating in the cafeteria, eating lunches out of little white boxes. Andy's kids are all presenting in the third and final orals session after the lunch break. William is sitting in his navy blue blazer and khakis, eating a side of macaroni salad out of a clear plastic condiment cup. He's saying little and appears fatigued. He readily admits he's tired, since he did his PowerPoint presentation the night before, finishing around ten thirty.

"I just want to get it over with," he half moans.

Shobhi looks lovely in a black sheath and blazer. Unlike some kids, who can appear as if they're playing dress-up when they don such adult apparel, Shobhi looks like a junior executive on the verge of promotion. She's penciled on some black eyeliner and smoothed down her hair.

But she's visibly nervous, talking at a comically fast clip, like a Looney Tunes character on amphetamines. Her left knee is doing that unstoppable giddyup.

Prem tries to get her psyched. "You'll do very well based on what I've seen," he tells her. "You're a powerhouse."

Connor has scrounged up a slick gray suit that has made the Li family rounds. At this point, he's not even sure of the original owner. But he feels like his attire is a good, confidence-boosting step up from his ill-fitting CT STEM blazer.

Andy's group says the morning presenters were very good, and they're particularly animated when describing a riveting project by a girl who came up with a novel way to determine the sex of dinosaurs. This is apparently very tricky to do using fossils. Who knew?

Andy has a policy of not watching any kids present but his. "I used to do it and it would just make me crazy," he admits. It often led him to feeling his kids were deprived of well-deserved prizes, and he'd drive home gripping his steering wheel, hugely pissed off on behalf of his team. So he's decided (with some encouragement from Tommasina) that the best way to be zen about the whole thing is to just not watch. Like someone who averts his eyes during the scary parts of a movie, he says, "I don't want to know, I don't want to see."

But when his kids are up, he's right there, armed with his Nikon and his new telephoto lens. He likes to document the whole process so the kids have a memento from the day, but it's clear that Mr. Busy Pants also needs something to do.

At 1:25 P.M., the kids head to the auditorium, a slick rotunda with three enormous curved screens positioned around the circular space, so the presenters can see their PowerPoint on the big screen with ease and no need to look behind them.

Connor is up first. He opens his mouth and starts speaking so quickly, it startles you into attention.

He begins by delivering the title of his project, which is "Fabrication of a Magnetically Vertical-Aligned Boron Nitride Nanotube Membrane in a Lyotropic Precursor for Water Transport Applications."

Connor then says, "And before I begin, I'd like to thank JSHS for this opportunity. It's quite unique and I'm quite nervous right now," which earns some lighthearted laughter from the audience. He doesn't ever fully shed his nervous energy, but that energy is dwarfed by his eye-popping mastery of both his project and the many scientific nuances involved.

Connor created an improved water filtration membrane that could be scaled up for industrial uses for things like desalination (removing salt from water). Its value-add feature is that the process simultaneously produces energy. Desert countries rely heavily on desalination; in Israel, roughly 55 percent of domestic water is desalinated.

Connor built his water filtration system out of boron nitride nanotubes, known to have water flow properties superior to their carbon nanotube cousins. When used to resalinate water (putting the salt back in) they can generate electricity, an added bonus.

During his presentation, he explains that he actually built a membrane by cooking the nanotubes onto a polymer. He did all of this junior year, but this year, he extended his project by creating 3-D video simulations that show how it all works.

He wraps up his presentation at the same warp speed he started and then the judges open fire, asking about everything from cost to where Connor obtained the nanotubes to whether he'd accounted for something called "membrane fouling," which is essentially the pores getting clogged during use and degrading the membrane's performance. Like an agile tennis player, he backhands his responses, sometimes saying, "That's a great question."

One judge asks, "Could you speculate if you doped the boron nitride if that would make it more electrically active?"

"That actually is an interesting point of consideration because research has shown that depending on the charge of the nanotubes themselves, the water transport properties changed. I believe the lower the charge is, the faster water can be transported through the nanotubes. So that's an important consideration."

The judge is satisfied. "Thank you very much." The audience applauds and Connor can finally exhale.

The floor is now open to the audience, and a kid asks Connor an astute question.

"I hate it when kids ask questions," whispers Andy. "It's a beef of mine. I'm gonna flatten that kid's tires on the way out. They're trying to sandbag them. It's bullshit."

As a former scientist and teacher, he would never ask a kid a question in a public forum. He sees it as deeply inappropriate. Let the judges do their job and keep your mouth shut.

The next presenter developed a chip for Zika detection. Rahul and his father, Prem, are interested to hear this presentation because it obviously overlaps with Rahul's Zika detection project. The girl presenting explains that current methods for Zika detection are time-consuming and require large samples. She's developed a system using nanoribbon technology. The benefit, she says, is that much smaller samples can be used to produce more sensitive and faster results.

At the conclusion of her presentation and questions from the judges, the floor is once again opened.

Prem, Rahul's dad, raises his hand. His question is both broad and highly technical. He essentially asks the girl to justify her design given that a method of diagnosis called PCR (polymerase chain reaction) has become the standard way to diagnose Zika, which undercuts the entire premise of her project. It's not an invalid question, but it's being posed by an Ivy League Ph.D. researcher to a high school senior.

In trying to answer the question, the girl fumbles. When she finishes answering, Prem goes in for the follow-up. His tone is testy, and he again pushes his central argument, which questions why her project has any relevance at all. The girl does her best to try to hop skip through this exchange.

Andy looks like there's a siren whirring in his head and like he might throw up. Prem's taking aim at this kid goes against Andy's grain. This girl, like many of Andy's kids, is a long-standing Connecticut science fair competitor, someone Andy finds to be very sweet and a good sport. But even if she weren't, Andy thinks adults have no business trying to take down a kid at a science fair. He's fuming, but he's also a conflict-avoidant person, so he says, simply, that today is not the day but that there will be a time and a place where he confronts Prem.

Next up is Shobhi. She walks up to the lectern and the minute she starts talking, everyone's captivated. It's like watching Nadia Comaneci in the 1976 Summer Olympics, where she was awarded seven perfect tens—the first such score in the history of the sport. Shobhi has curated her talk to perfection. She has anticipated every question imaginable and answers them before they can even rise to your consciousness.

Shobhi used publicly available mass spectrometry blood sample data of mice who were engineered with human cells to develop pancreatic cancer and compared them against controls, or mice free of pancreatic cancer. She then built an algorithm to analyze all the protein fragments to determine which are the key players in pancreatic cancer and which ones don't appear to have meaning. Previous work in this area has achieved accuracy of about 60 percent. Using the data available to her, Shobhi got to 80 percent—largely because she combined two statistical methods that allowed for more precise and nuanced results.

She's such a fluid talker and also very relatable. Computational biology, the category in which her work falls, is beginning to gain a lot of traction, but it's still a new area for high school kids, in large part because it requires substantial knowledge of coding—which is not standardly taught in schools. Shobhi taught herself a great deal and later attended some summer camps that offered coding classes.

At the conclusion of her presentation, she says, "Finally, I implemented my model in a rapid, user-friendly web application, to show how it would be used in a clinical setting. Overall, this research opens a new pathway to accurate blood-based detection of premalignant pancreatic cancer, which would allow doctors to treat it while curative surgery is still possible.

"Uh, am I out of time, or . . . ? Okay, I'll make it fast then," she says, quickly delving into areas of future exploration stemming from her project.

She fields a few questions, but her talk was so airtight, there frankly isn't a whole lot to ask.

Finally, William is up.

"The last speaker is William Yin," intones a judge.

William opens with, "So, my name is William Yin and my research involves the development of a portable, low-cost, tattoo-based biosensor for the noninvasive self-diagnosis and quantification of atherosclerosis."

He then defines atherosclerosis ("the formation of lesions within the lumen of the artery that then leads up to the buildup of plaque deposits"). He talks about its contribution to heart attacks and strokes, its underdiagnosis and its prevalence.

Like Shobhi's, William's manner of speaking is very relatable. He also has a slight pitchman quality, a smooth salesman at work. You almost expect him to come out with "But wait! There's more!"

In William's design, you'd place a little sticker on your neck and press a button that sends nanoparticles into your carotid artery using a mild electrical current. A second little jolt sends the nanoparticles back up to the surface of the skin. He's devised measurements that tell you whether plaque is present based on the amount of nanoparticles that come back, the theory being that if you have atherosclerosis, a critical mass of nanoparticles will hang back and stick to the plaque.

He concludes his speech with an Oscar-length list of thank-yous, seeming to acknowledge anyone who's ever said the word "atherosclerosis." He even thanks the guy who owns the printing company that prints his posters, even though today he's using a PowerPoint. He concludes with a shout-out to the only person on his list who's in the room.

"And finally, my mentor, Mr. Andy Bramante, who stuck with me through the entire duration of this research. And thank you to all of you for taking the time out of your day to listen to our presentations."

Applause.

William then endures a rather brutal question-and-answer round from the judges. One judge is concerned that if patients were to place William's sticker on their necks with too much force, they could actually dislodge a plaque and *cause* a stroke. It's not an unreasonable question, but she's very coy in asking it, poking around to see if William catches her drift rather than just coming out and saying it. He makes a few runs at trying to field her shrouded questions. She repeatedly probes him about his decision to use the carotid artery as the site for the sticker and if he had consulted with actual cardiologists.

A second judge jumps in and though she concedes that William's overall design has some optimal features, she also questions its safety. He's rescued by the clock when a third judge announces, "That's all I have time for, so . . ."

"Yeah, that was bad," mutters Andy. His take is that you never argue with the judges—no good will come of that. But William's interrogation didn't feel so much like an argument, more like he was being hammered into a fast-acting pile of quicksand. It's simply unknown whether placing a biosensor tattoo on the neck in any way poses a hazard to patients. Remember, William demonstrated his concept in the lab but is not able to test it on humans or animals. So there are natural limits to what he can report.

This is a shining example of the unavoidable but rocky terrain of the inequities of judging. Today, several kids have been tossed the softball question of what sparked their interest in their topic. William has been asked to answer to one judge's perceived issue of whether his device could essentially cause serious damage to a person. She had her bone and she wasn't dropping it for anything.

Afterward, he's somewhat blasé about it, acknowledging that it wasn't the best of exchanges. He was thrown off kilter—neither he nor Andy had ever considered this issue—and never quite steadied himself.

With the presentations over, the kids file back into the cafeteria to await the awards ceremony. They've been here all day and look depleted, slumped in their chairs. When fair head Joy Erickson steps up to the podium, she mercifully spares the crowd any speeches.

"In fifth place, we have Lauren Low," announces Erickson. Lauren is the kid who developed the Zika diagnostic that attracted Prem's attention.

Fourth place goes to another girl, as does third.

"In second place, we have . . . William Yin, who will travel to nationals and receive a fifteen-hundred-dollar scholarship for college." William stands up with a bewildered look on his face and heads to the podium. His mom, Wendy, has shown up and is beaming and clapping as William accepts his award.

"And in first place, we have Shobhita Sundaram, who will travel to nationals and receive a two-thousand-dollar scholarship," says Erickson.

Andy's clapping and looking rather amused. This is the first time he's ever had two kids take both spots for the JSHS nationals and he can't quite believe it's happened in this way. He's not surprised about Shobhi's win, but he's somewhat floored at William's, given the beating William took from the judges.

Rahul and Connor congratulate Shobhi and William, and the kids slowly make their way out of the building. It's dark by now and the wind is pressing a blade against any exposed skin.

Andy's beat too. Nearly falling into his car, he says, "I just want to get home." It was a high-stress day, with some unexpected twists. And yet Andy has no idea how much twistier and more stressful it's about to get.

The night of the competition, Prem sends a lengthy email to the fair heads. He writes: "I have to raise a strong objection to the selection of Lauren Low as (a). a [*sic*] oral speaker (b). as a finalist in light of the abject failure of her project, over that of my son Rahul Subramaniam's. My son was selected to present a poster, but Lauren was chosen for a [*sic*] oral presentation." He then writes a detailed analysis of the flaws in Lauren's Zika project, as well as a fourteen-point description of Rahul's project and the ways in which it was superior.

The email then takes a personal turn: "I was surprised to learn that Lauren Low is popular among the judges, and saw the judges interact with her after her talk. Is this why she was given a leg up?"

In searching for an explanation, he also raises the Greenwich issue, writing, "There has been a tendency for students from Greenwich High School, in Greenwich CT to be overlooked simply because they take many of the top honors . . . Since two oral speakers were already selected from Greenwich High School, did the judges make a conscious decision to NOT pick another GHS student? This is very unfair to students."

But in the end, he brings it back to Rahul. "What am I to tell my son?" he writes. "All his work does not count, but due to political reasons someone else was chosen in his place, and the person went on to win a finalist spot even though she performed only one experiment which also failed? They both wanted to detect Zika virus, but only one of them devised a completely unique system, and showed it worked. I am asking that my son's project be reviewed and weighed again and if Lauren Low is chosen, my son's work be considered superior to hers in all respects."

When the email goes unanswered, three days later Prem writes again, saying, "I would like to know how the situation is being remedied."

JSHS fair director Joy Erickson replies and says she consulted with the judges and the results are final. She writes, "Your attack specifically on Lauren Low was very inappropriate," and refuses to engage in any debate about the girl's project. She dispassionately details the allotment of spots and points out that every school in Connecticut receives the same number of potential presenters.

Prem insists on having the last word. In a subsequent email, he writes, "Based on your reply, you reviewers are the ones who stack the deck. This is not about how you apportion the applications and spots. This is about how you award. So no matter how bad a project is, or how shoddy it is, you have to give it a [sic] award so that the better ones don't all come from the same school. Really? What's merit about then? Many oral presentations were middle school standard.

"My disappointment was exacerbated by my witnessing the whole way in which this symposium seems to operate. Let's put my son's poster aside. Very frankly, I was astonished to find most oral presentations were utterly sub-standard to even other posters I read . . . I would not have let my son present such shoddy projects under any circumstances. How could you as the organizers."

As for Erickson's consulting with her judges, Prem writes, "Your conscience is what you should consult, on how your symposium manipulates merit versus politics."

He goes out with a bang, wrapping up the exchange with this:

I shall be warning all high school students who come to my lab at Columbia University or those of my colleagues to steer clear of JSHS, unless they are chosen for any oral presentation. Even then I will have to remind them this is a popularity contest. There goes our education system.

Sincerely,
Prem

Andy is copied on these exchanges and when he sees them, he cannot fathom what he's reading. Prem did not speak to him ahead of time, so there was no opportunity for Andy to counsel him against such vitriolic action.

Andy's also volcanic about being copied because he in no way wants the JSHS folks to think he's complicit or supportive of Prem's positions. It's bad enough that he's constantly fighting the taint of Greenwich, the widely held view that it's a place populated with snotty, entitled, wealthy families who somehow float above the rules by which the rest of society is bound. He's grown so sick over the years of the snide comments and assumptions that his kids start out with advantages that somehow dilute the meaning of their work and victories.

In Andy's opinion, for a parent to craft such a condescending and vituperative email to fair heads completely validates every snicker, every aside, every dig. Because in the face of this behavior, how can he argue anything to the contrary?

He's not sure how exactly to confront Prem. It was bad enough when the scientist tried to take down the Zika kid during the question-and-answer period at the competition. But this is unforgivable. Ultimately, though, he never says or writes anything to Prem. This is an example of where the inherent good guy in Andy, mixed with his need to be liked and further exacerbated by his extreme conflict avoidance, leads to nothing but him grousing about Prem's indefensible behavior—but not doing a damn thing about it.

As for why he's allowed Prem in the classroom to supervise Rahul's work in the first place, until now, Andy's seen him as potentially helpful. Prem has always been willing to field questions from kids working on biology-based projects, which aren't Andy's specific area of expertise. Prem's generous with his time, but his primary purpose is to mentor Rahul. Rahul even officially lists his father as his mentor on most of his competition applications this year.

Until now, it's never been apparent to Andy how invested and serious Prem is about Rahul's work and success. Although maybe it should have been: when talking about Rahul's project, Prem uses the first-person plural "we," saying in the beginning of the year, "We'll see what we can come up with."

Most parents of kids in the science research domain will feel the burn of defeat. There will always be a small group of winners relative to the entire pool of entrants.

But this is high school. These are kids. It's incumbent upon the parents to do all the things we all know about, in theory, when it comes to loss—and pass that along to their offspring. Be gracious in defeat, know it's better to enjoy the process than the outcome, trying and losing is better than not trying—all that.

In cases like this, you're left to wonder how and when did the stakes become so artificially inflated? Despite all the admirable love and devotion parents in this town bestow upon their kids, one thing is rather clear. When it comes to their children, parents here (and elsewhere, of course) are guilty of inability to take the long view. They assign gargantuan, disproportionate importance to every incremental thing, so that every quiz, every test, every teacher, every game, every score, every project, every music composition *matters*. Much of it seems to be tied up in the college arms race, since a good portion of society has concluded that your total worth as an individual depends on where you go to college.

But when you're out of college and a bit further along in life and you look in the rearview mirror, how does your high school experience rate, in terms of overall importance? Do all those contests and quizzes and games inform wide swaths of your daily life now? Chances are they don't. Yet even knowing this doesn't seem to guard against the gripping obsession some parents have with their kids' achievement.

While these science competitions have become so sought-after, Andy has tried to keep everyone's perspective in check, to remind them that he wants the kids, above all, to learn and have fun.

"I always tell the parents, 'The shingle outside the door doesn't say 'science fair.'"

41

Ethan

The forty Regeneron STS finalists are invited to Washington, D.C., the week of March 9–15. This year, there are twenty-five boys and fifteen girls from thirty-four schools in seventeen states. They were selected from more than seventeen hundred applicants, nearly all of whom had likely won some impressive awards at local or regional competitions. In other words, these forty kids are la crème de la crème de la crème.

Regeneron STS says the competition "focuses on identifying the next generation of scientists and engineers who will provide critical leadership in solving some of the world's most pressing challenges while shaping the future of research and development for our nation and the world."

Reading the bios of the finalists, some themes emerge. Many of them are distinguished musicians, play sports, and do some kind of volunteerism. Twenty-seven of the forty are Asian—fourteen of South Asian descent and thirteen of East Asian descent. Most attend public schools.

This year, Regeneron's first as the title sponsor, is particularly resonant because the drug company's cofounders, Drs. Len Schleifer and George Yancopoulos, have their own personal history with STS, having advanced far in the competition back when they were high school seniors.

The scientists take their sponsorship seriously, which is evident from their ten-year, $100 million commitment to the Society for Science & the Public, which oversees STS, ISEF, and other programs. But perhaps even more telling is that Regeneron has significantly upped the prize money for winners—nearly doubling the overall award distribution to $3.1 million annually, making the top prize $250,000 and doubling the semifinalist awards to the kids and their schools to $2,000.

Regeneron has a sprawling corporate campus in Tarrytown, New York, which houses its drug research and development facilities. Sitting in his conference room, George wears hiking pants and a casual short-sleeved shirt. His rugged look suggests he could be going surfing and offsets the instinct to address him as "Doctor." A youthful, scruffily handsome fifty-eight-year-old, he's got the self-assurance of the high-net-worth individual he is (in 2014, he was estimated to be worth $400 million; that same year, Len reportedly hit billionaire status), but with an earthy ease about him.

It's February 27, two weeks before George and Len will go to Washington to head up Regeneron's inaugural STS awards gala. George is eager to talk about why a $35 billion biopharmaceutical company would take hold of a high school science competition.

"The [Regeneron] board of directors sort of turned around on me and basically said you've got to do this because it's the right thing to do for this national treasure," he says. "Because we believe so strongly that we are the best science-driven corporation in America, it shouldn't be in anyone else's hands. We can do the best job of helping continue to nurture and grow it."

As the first-generation son of Greek immigrants (his father's first job when he arrived in Queens, New York, was sewing fur coats), George's ticket to greater things was attending New York's famed

Bronx High School of Science and becoming a Westinghouse STS finalist back when that company was the fair's title sponsor. Today, he remains bullish on science and why we all need to maintain some perspective on its value.

"Who's going to save us? Whether you're talking about cancer, whether you're talking about Ebola, any other epidemic that might come out of nowhere. Now we've got to worry about bioterrorism, let alone climate change and the environment, new forms of energy. Who's going to do that? It's not going to be the hedge fund managers who do this. It's going to be some brilliant young kids who are attracted to science and recognize that they might have the ability to solve the biggest problems facing mankind," he says, just getting started.

"People talk about the war on this or the war on this, this is the biggest battle mankind has to fight. We have to win this war against disease, against aging, against all of these obesity-related epidemics, against climate change, against limited resources, everything that's changing about our world.

"I mean, if we solve everything else . . . if we win the war against terrorism and cyberwarfare and all this, we've still got to solve all these problems and who's going to do it? I think the best that we can do is to make sure the brightest, sharpest young minds get attracted to science and fall in love with it and realize the power of what they could do, how they could really change the world. It's not like the world will be *better* for it. We are absolutely one hundred percent *dependent* on it."

Sunday, March 12, is public viewing day, where people come to meet the STS finalists and hear about their work. Earnest parents walk around with their young children in tow, as if they'll be meeting celebrities. Many local scientists attend, as well as former STS finalists. And television crews circulate, looking to snag interviews with the junior genius set.

This year, the public viewing is being held at the National Geographic building. Each of the finalists is given a table where they display their poster, as well as a little stack of colored cards with their bio to distribute to interested parties.

Ethan and Derek are both wearing sharp sport coats and looking eager for takers. This will fade significantly as the afternoon wears on and they've launched into a description of their work for the thirtieth time.

Derek, a keen social observer, is enjoying himself but has noticed a certain homogeneity among his finalist peers. "Everyone has the same personality," he says. "It's kind of creepy."

Before today, the finalists have been spending nearly every waking minute together when they're not being interviewed by judges and put through rigorous mental exercises. They go individually before a set of judges and are given random science questions, ranging from "How much voltage is in this light bulb?" to "What is graphite?" to "Why can't Godzilla and King Kong exist on the same island?" Sometimes the questions are super technical, like "Why is there frost on the roof of a car when the temperature outside has never dropped below thirty-eight degrees Fahrenheit?" Others are bigger picture, like "How do you extend human life?" A favorite among the finalists is "If you were to design a dragon, how would you go about doing it?"

The judges seem to be looking well beyond the kids' work, assessing the quality of their thinking and how creative and dynamic they can be.

Ethan relishes these kinds of challenges, and given the scope of his scientific interests and endeavors over the years, he's well armed to give good answers.

"I really enjoyed those. These are really cool problems they have you solve," he says, adding, "I wasn't as up-to-date on some of the bio ones."

It's impossible not to be gobsmacked by the sheer brainpower in the room, as well as the often humble and quiet ways these kids have toiled away on their projects, some for years.

A particularly impressive finalist is Amber Zoe Yang, a seventeen-year-old from Florida. A petite girl with a sunflower of a face, here's how Amber explains her project: "Essentially what I'm doing is using machine learning to identify and track and predict future orbits of space debris. There are five hundred thousand space debris samples that are orbiting in space right now and they pose imminent dangers to things like the International Space Station. So if I'm able to predict a space debris sample in enough time in advance into the future, I can warn things like the ISS [International Space Station] that the collision could be in its future so it could move out of harm's way."

How cool is that? In space, there's a giant floating junkyard containing an estimated half a million pieces of debris (old satellites, for example) ready to collide, and Amber has come up with a machine-learning method to track and prevent such collisions, which can damage satellites and spacecraft. The current method uses a static statistical model and software that requires manual updates. Amber developed two artificial neural networks (ANNs), which are computer-based models that simulate brain circuitry and patterns. Her ANNs are able to apply Kepler's laws of planetary motion to the orbiting space debris and give regular updates.

Regeneron's finalist booklet says her work "may be a step toward safer space travel amid rising populations of space debris." A small contribution, really.

Derek's parents, Charles Woo and Ryeo-Jin Kang, have invited Andy as their guest to the Regeneron awards ceremony, which happens to coincide with a blizzard in the Northeast. Charles and Ryeo-Jin drive down a day before the snow hits. The Noveks made a trip out of it and have been in town all week.

But since Andy would need to travel down the day of the awards, he decides to pass. "My motto . . . no train rides in a blizzard," he texts. In the end, he doesn't have to agonize over his decision because Amtrak cancels his train.

It's a shame he won't be at the big awards gala, as this is the first time he's ever had two finalists in the same year, and he's so fond of both Ethan and Derek. The good news is that the Society for Science & the Public, which oversees the competition, live-streams the awarding of the top prizes.

So on the evening of March 14, the finalists and their parents and teachers (like Andy, some have been waylaid by the weather) convene at the National Building Museum for the big evening.

It's one hell of an event—these kids are made to feel like the rock stars they are. STS rents tuxedos for the boys and offers the girls either hair styling or makeup. Nearly all of the latter wear floor-length evening gowns. Dinner will be served in the main atrium of the museum at tables adorned with luscious floral centerpieces. Giant screens have been mounted on all sides so everyone can see the speakers and kids.

During the cocktail hour, the kids again stand in designated spots and chitchat with guests. But there's a jittery current in the room because it's Pi Day, 3/14, the day when MIT notifies applicants whether they've been admitted. As an added nerdy touch, the outcomes are sent at 6:28 P.M., or Tau Time.

Derek is at his spot and at 6:28 he checks his phone. All of a sudden, his face illuminates into a smile so wide, it threatens to split apart the walls. He finds his father, and when Charles hears the news, he bursts into tears. All of a sudden, this six-foot-tall Korean-American giant is weeping and wiping his eyes with his fingers. After all the agony and heartbreak following his initial MIT deferral, his boy's done it. Charles immediately calls his parents, who live for their grandchildren and have kept apace of the whole college process. As he delivers the news, his voice is high-pitched; he's again on the brink of tears.

This may sound extreme, but when you're the child of immigrants who sacrificed unspeakable amounts to give you—and then your kids—a solid education and a chance to prosper, an elite college acceptance is a golden dividend.

Several of the other kids also learn they've been accepted to MIT.

But there's no time to dwell on it because the STS festivities are about to begin.

Guests are led to a small room for a medaling ceremony where the kids are called up one at a time and have gold medals placed around their necks, like Luke and Han at the end of the original *Star Wars*. Dr. Len Schleifer, the CEO of Regeneron, makes a few remarks and gets some laughs when he mentions that back in the day he was a mere semifinalist, while George Yancopoulos was a finalist.

Following the medaling ceremony, everyone moves to the main atrium for dinner and the top awards. The kids don't sit with their parents but are placed at tables with distinguished guests.

The keynote speaker is Dr. Angela Duckworth, the psychologist known for her work in the area of grit (and author of the bestselling book of that title), the trait that is the result of passion and perseverance for a singularly important goal. According to Duckworth, grit is the hallmark of high achievers in every domain—and the good news is that it can develop and grow. It's not a static trait. She's a compelling speaker, even if the message smacks a bit of preaching to the choir. None of the forty kids assembled in this room appear to be lacking in grit.

There's a fun video montage of the finalists during their week in D.C. And each year, the kids select a fellow finalist to deliver a speech. This year, the honor goes to Jackson Barker Weaver, a charismatic Harvard-bound kid (he was accepted early and is also a third-generation Harvard legacy) who examined the use of electricity to increase the kinetic energy of molecules, which may have implications in the use of electrical currents in medicine.

Jackson's speech strikes the perfect balance of cheeky humor and inspiration. He talks about the dragon design question posed to the finalists.

"Personally, my dragon had four key parts," he says. "It had the legs of an ostrich, covered in the shell of an armadillo for protection. It had a heavy tail to maximize torque when it swung. It had the skeletal wing structure of a bat, which is strong and sturdy. And finally, the most important feature, it had a fire-breathing fuel seeping from its gums, ignited by the heat of its breath.

"I felt listening to other answers in comparison to mine of the many wonderful finalists that the future of dragon genetics in this country is stronger than ever," he adds, earning a big laugh from the crowd.

Jackson talks about how the greatest bonds among the finalists were formed over their deficiencies, the things they don't know, and that in many ways, the week was a healthy reminder of all that remains to be learned. He speaks without a trace of false modesty about the importance of remaining humble. It's a winning performance.

And then the moment everyone's waiting for, the awarding of the top cash prizes. All of the finalists have won $27,000 for making it this far. The top ten will go home with $40,000 to $250,000.

The top winners are announced by the judging chair, Dr. Sudarshan Chawathe, a computer scientist and professor at the University of Maine.

Dr. Chawathe says, "I was told not to open the wrong envelope," a nod to the recent Oscars gaffe when Faye Dunaway and Warren Beatty were handed the wrong envelope and incorrectly named the winner for best picture. He spares the anxious crowd a long speech and gets right to the task at hand.

"The tenth-place prizewinner and the recipient of a forty-thousand-dollar award, from Alexander W. Dreyfoos Jr. School of the Arts in West Palm Beach, Florida, Stefan Wan." Music plays and Stefan makes his way to the podium to receive his award.

"The ninth-place winner and the recipient of a fifty-thousand-dollar award, from Lake Highland Preparatory School in Orlando, Florida, Vrinda Madan." Vrinda's parents are seated at the same table as Bonnie and Keith. Her mother lets out a shriek of delight and she and her husband start clapping. The other parents at the table offer hearty congrats.

"The eighth-place winner and the recipient of a sixty-thousand-dollar award, from Greenwich High School in Greenwich, Connecticut, Ethan Joseph Novek."

With his $2,000 semifinalist prize and now $60,000, Ethan can cover most of his first-year Yale tuition, which is currently $70,570. Given his parents' outlay for the construction of his carbon capture prototype, this is not insignificant.

Andy's watching the live stream from home and taking screenshots of Ethan as he heads to the podium with his name and face projected around the room.

However, he and others who know Ethan well are actually a tad surprised that Ethan isn't farther up the victory ladder. But who can feel glum about placing in the top ten and winning $62,000? (Ethan later says that the finalists had a betting pool on the top winner and many of them had wagered on him.)

The parents of the third-place winner, Arjun Ramani, from West Lafayette, Indiana, also happen to be seated at Bonnie and Keith's table. So everyone's joyous and can display it unabashedly since they're all parents of tonight's superstars.

The top prize of $250,000 goes to a girl from Oradell, New Jersey, Indrani Das, for her development of a possible treatment for brain injury. Given the increasing public fear of concussions and traumatic brain injuries, it's a timely project. Four of the top ten winners are girls, which is impressive considering there are only fifteen female finalists.

When Indrani's name is called, balloons and confetti rain down from the ceiling. She looks completely bewildered.

42

Sophia

When Sophia arrives at CSEF on March 14, she immediately finds her designated spot and carefully sets up her poster and materials. She stands back to look at everything, takes it all in, and makes some adjustments. And then, like all the other competitors, she goes home and waits.

The very first round of judging at CSEF involves no presentations by the kids. Instead, the judges circulate, read the posters, and select finalists, who will present the following day.

Back home in Old Greenwich, Sophia is extremely nervous. As is the case for anyone working on something for so long—it's now been a year and a half—her perspective on her own work is murky. She's so enveloped in every nuance of this project, she has no ability to comment on whether it's good, bad, or other. And the kids tend to shy away from too much analysis because in their minds, all the power lies in the hands of the judges. So there's no use patting themselves on the back.

CSEF has said it will post its results at five P.M., but due to the inclement weather, results will not be made public until seven P.M.

Sophia's going to have to wait, anyway, because she has a tennis lesson tonight.

When she walks in the door from her lesson, she drops her bags and heads straight to her laptop. She goes to the CSEF website and searches for her name.

"Chow."

There it is. She's a finalist. She's going to get her shot.

Sophia's got other things on her mind these days besides her Lyme project. The junior prom is just a month away.

Almost bigger than the prom itself these days is the promposal, the manner in which someone asks for the hand of his or her date. A promposal can range from a poster with a clever pun (popular) to a flash mob (elaborate) to all kinds of original, zany things. It's not an exaggeration to say that in some cases the amount of planning and agonizing and staging surpasses many wedding proposals.

And because this is high school, no one wants to ask someone who doesn't want to be asked. So people often pre-ask or pre-book their date and then prompose.

Many of the girls have been dreaming of this day all year. Some bought their prom dresses in the fall. There's a Facebook page where GHS girls post photos of their dresses, not for bragging rights, but to try to guard against duplicates. No one wants to show up at the prom only to face her dress doppelgänger.

Sophia and her friends, working like efficient honeybees, broker dates between themselves and their popular male counterparts. In some cases, these deals are made easy by the fact that there are already established couples. When there aren't, the girls make the matches and the boys almost always agree, since it spares them from having to put much thought into the matter.

But Sophia wants to go with a guy outside of her social network, someone in one of her classes. This means that her girls will have to relay this fact to him and gauge his interest. We'll call him Taylor. He's cute and plays baseball. Like Sophia, he's shy.

Shortly after CSEF, the deal is made. Sophia and Taylor are going to go to prom together. This has been acknowledged and confirmed between the two of them. (And they didn't even need to get lawyers involved.)

A couple of weeks later, Sophia and her friends are hanging out on a dock in Old Greenwich. Her friends seem unusually amped up.

All of a sudden, Taylor shows up, colorful bouquet of flowers in hand. He's made a baseball themed poster that says, "Can I steal you to prom?"

Sophia, of course, says yes, and her friends snap a lovely picture of the prom-bound couple, arms around each other, Sophia clutching her bouquet. The photo is taken just before dusk, what Hollywood people call "the magic hour," when the natural light has a slightly burnt-orange tint that makes nearly everything and everyone look glorious. She's extremely happy and also seems relieved, as if there were a real chance Taylor might not want to go to prom with one of the most stunning girls at Greenwich High School. It's a winning and genuine trait of Sophia's—the complete oblivion that she struck gold in the beauty gene lottery. She spends much more time looking at her biofilm samples than at herself in the mirror.

No one but her is surprised that she got the guy.

43.

- - - - - - - - - - - - - - - - -

Andy

At 7:26 P.M. on the day CSEF announces its finalists, Andy's got his laptop open and is repeatedly trying to get www.ctscience fair.org to load. But he's having some technical difficulties, which inserts an unnecessary level of agony into the reveal of the results.

He gets a text from one of his students saying he's not a finalist, which instantly bums him out. It's from one of the sweetest and most humble ones in the pack. One consolation is that this kid was one of the few science research seniors who was accepted early to his first-choice college. He's headed to Duke in the fall.

Finally, after several restarts and refreshes, Andy's able to view the list of finalists. It's alphabetical, not by school, so he has to scroll through the entire list. He's reading off the names and then instantly deflates.

"Ugh." Another deserving kid, a sophomore, didn't make it. By the time he gets to "Z," Andy sees that seventeen out of his nineteen kids have been named finalists, an astonishing feat. But instead of feeling gleeful, he's morose about the two who weren't chosen.

Collin, the sophomore, is a mix of quirk, genius, and likable packaged in someone who this year is turning the corner from child to teen. When he opens his mouth to discuss science, he dives right into the meat without any windup or intro. As he talks, light bounces off the glint of his considerable orthodontia.

He started out that year by saying he's really been born into the wrong era, that he would have loved to have been part of civilization-altering engineering feats, like electricity. But since he missed that boat, cellular engineering is the next best thing. Here's a brief excerpt from his abstract:

> The purpose of this research was to develop a system which can be easily applied and not only mirrors, but improves, the versatility of deoxyribozymes. The system is based on the RNA-Ligating 9DB1-deoxyribozyme, due to its stability and functionality in conditions similar to that of the cytoplasm. The purpose of this system is to perform an if-statement where the inputs are triggered by up to two specific sequences of RNA in solution and the output leads to the synthesis of any chosen protein.

Best to stop there because likely fewer than ten people reading this will have any idea what it means. The biggest-picture explanation is that he showed that RNA can be manipulated in such a way that you could, say, selectively kill cancer cells. It's as if he's training the RNA to be cancer assassins. His methodology and execution were so high-level, it took Andy a good deal of time to fully grasp what Collin was proposing.

It's clear within about a nanosecond of speaking to this kid that he thinks at an altitude the rest of us only ever reach while in an airplane. Following a particularly brutal Sunday work session with Andy that stretched past ten P.M., Collin's project eventually worked. (It involved several long wait times for chemical reactions to occur.) Andy had high hopes. For sophistication alone, it stands apart.

So as is his way, despite the feat of having seventeen kids make it to the finalist stage, Andy's perseverating about the two who don't make it, worrying how they'll take it and how he'll frame his pep talk.

Before tonight, he and Collin had had a recent exchange via text in which Andy wrote, "You're just getting started, my friend. No one said it would be easy. And, btw, you are enjoying the class, no?" Collin replied, saying that Andy's class was the best he's taken and he was having a great time. So there's that.

And since Collin is still a sophomore, many opportunities await him. Andy reminds him of Shobhi's CSEF snub last year and highlights the fact that she still went on to advance in several prestigious competitions.

Despite his disappointment, the next day is game on at the fair and he has to set all of it aside and focus on the kids. Not that there's much for him to do, especially this year, when he's been ordered to remain out of the gymnasium while the kids are being judged. In prior years, CSEF allowed him to be the fair photographer of all finalists. But there's been a growing sense that he shouldn't be near his kids during the judging. Because he and his kids have dominated the event, causing no small amount of protest over the years, the fair heads don't want to give other people any unnecessary ammunition. Andy thinks the "Do not cross" tape is petty and juvenile, but he, of course, abides.

The fair is held at Quinnipiac University, a private school in Connecticut that gets a good deal of publicity producing political polls during elections. Under Sandy Müller's leadership, CSEF runs with the efficiency of a well-fitted steam engine. There is signage everywhere, the judges are organized and fed, there are legions of volunteers on-site to help shepherd the kids and get them what they need. You can't walk five feet without encountering someone who can answer any question you may have.

The judges are all volunteers. Sandy and her staff enlist them from local science, engineering, technology, and pharmaceutical companies, as well as colleges. They usually take the process very seriously, knowing that the CSEF stakes are important—the top winners go to Intel and ISWEEEP and win valuable scholarship money.

The kids and their posters are set up in long rows in the gymnasium. They stand at their spot, anxiously awaiting the judges, often making small talk with one another. Each entrant will be seen by a minimum of three judges and as many as nine. Sometimes the judges are alone and sometimes they travel in small groups.

Ethan arrived back from Regeneron STS the night before, but he turns up at CSEF, much to the delight of several of the judges and fair heads. Ethan started competing in eighth grade and quickly became a fan favorite. He didn't come to CSEF during his junior year because he was working at Yale and hustling to get his peer-reviewed paper published. His absence was felt.

"The judges love Ethan, oh, my god," says Andy. "It's like their son is coming home."

Sophia's on the end of a row by herself, apart from a few clusters of GHS kids situated elsewhere throughout the gym. Her Lyme disease is dragging her down a bit. She looks crisp and smart in her navy blue blazer and low heels, but her eyes are sleepy, ringed by dark circles. By the afternoon, she's sitting down, looking as if she might melt into the metal folding chair and take a nap.

However, when a judge comes by for her presentation, she jumps up. It's as if a wilting flower has been put in a vase and set down in the sunlight. Her eyes sparkle with the adrenaline of the moment.

Some judges are friendly and try to put the kids at ease. Some are all business and aren't interested in any niceties or small talk. Others make the judging more of a conversation. This judge cuts right to the chase.

"Can you give me an overview of what you've done?"

Sophia, almost always a reticent speaker, launches into a smooth introduction of her work.

"So my project is titled 'PDE4 Inhibitor-Assisted Antibiotic Suppression of *Borrelia* Biofilm Growth for Treatment of Chronic Lyme.' So what I noticed is in Lyme disease, a lot of times patients rely on antibiotics for treatment and I wanted to make those more effective. I wanted to make antibiotics the most effective that they could be."

She then talks about PDE4 inhibitors and prior research showing their effectiveness in removing biofilm that had grown on medical devices implanted in the human body—and that she set out to see if PDE4 inhibitors alone and in conjunction with antibiotics could remove the biofilm that surrounds Lyme disease bacteria in the body.

"So the PDE4 inhibitors that I chose were rolipram, roflumilast, and cilomilast. And rolipram is mainly used for antitumor suppression and the roflumilast and cilomilast are used in airway diseases of inflammation, so I thought maybe that could be applicable towards Lyme disease or infectious diseases with inflammation," she explains. She then gestures to her poster, one of the snazzier-looking ones at the fair due to her 3-D graphs and color coding for all the combinations of drugs she tried.

"This is the control image. This is just biofilm with nothing on it, and here is rolipram and roflumilast and cilomilast, and as you can see from the pictures, it's very evident that something happened to the biofilm. So in order to get the info that I needed, I took an image analysis . . . So rolipram was the most effective with around thirty-one percent, roflumilast twenty percent, and cilomilast around two percent. That allowed me to go back and add antibiotics because I knew they'd be successful."

She then describes the cool way she empirically measured the reduction in biofilm—by measuring how much light it scattered or didn't. She cites the paper she found that used this methodology but tells the judge that her work produced very different findings.

"They actually proved that as biofilm grew, reflectance decreased and the light scattered. So my result indicated the opposite of that. I used a fishtail optic and I don't want to discredit what they proved because it might be different, but my results indicated that as biofilm increased, so did the reflectance. So that just reiterates what this picture says, as biofilm increases, reflectance increases as well."

Sophia conspicuously does not mention that she has Lyme disease. She has decided that unless explicitly asked about the genesis of her interest in the topic, she won't mention it—despite the fact that it could win her some sympathy. But that's not how she plays the game.

It's fascinating to see Sophia, normally so shy and reserved, have such a commanding presence when describing her project. It's a whole new her. Standing tall, she uses all her visual aids and she's even brought the fishtail optic device for the judges to see.

The narrative of her work has a nice, logical progression. She doesn't get into the difficulty she experienced in growing the biofilm itself.

The judge, all business, says, "So, the real discovery then was using this particular device to measure the reflectance . . . discovery of the correlation between reflectance and growth?"

The judge needs a little prod from Sophia to fully grasp the scope of her work.

"It was also to enhance the effectiveness of antibiotics and showed that if you add PDE4 inhibitors to antibiotics it's going to be successful in reducing the biofilm. And previous literature suggests that antibiotics are eighty to ninety percent successful if there was no such thing as biofilm covering the bacteria," adds Sophia.

Satisfied, he replies, "So yeah, so it's that as well. So there's two pieces. You discovered something in addition. So what's next for you?"

"Well, I want to work with the real *Borrelia burgdorferi* at, like, a higher-level lab."

"So you're an eleventh grader. Are you going to continue to do work on this?" he asks.

"If I can find a lab that will let me," replies Sophia.

"I mean, I'm sure there are some. Where are you? You're at Greenwich? You can talk to Andy," says the judge.

"Yeah, he's my mentor for this."

"So he can connect you with someone that would allow you to use real . . . ?"

"*Borrelia burgdorferi.*"

"Right," says the judge. "That's awesome. Great work! Very nicely done."

Once the judge walks away, Sophia smiles and lets out a big sigh. It's been a good, if exhausting, day, one where it seemed like her Lyme might rob her of her moment. But she rallied and now she can finally go home and collapse.

By the middle of the afternoon, Sophia isn't the only kid who's looking somewhat sapped; they all are. The stump speeches get a little wilted with each round. Before the judges walk away and head to their next kid, they'll occasionally toss a compliment to a presenter. Very rarely, a judge will be so taken by a project, he or she will let the wow factor be known. But mostly they thank the kids, immediately start jotting notes on their clipboards, and leave.

Overall, most of the kids are pleased. Rahul had been concerned that the judges would penalize him for not working with live Zika virus, which he wasn't able to do in Andy's lab, the same way Sophia couldn't work with pathogenic Lyme bacteria.

"It was something I was worried about, like, oh, you should have worked in an institution with live Zika virus and full RNA," Rahul says. "But they've been very understanding of that, so that's a real relief."

Shobhi doesn't appear to have to worry about the judges not understanding her project this year. It seems the fair had plenty of people on hand with the appropriate knowledge base. Since Shobhi already won an Intel spot at the CT STEM Fair, she's here to compete for specialty award money—and also to hone her presentation.

"It was a change to present to people who knew more about the computational side than the biology side," she says. "It was really nice and there was a lot of stuff I didn't have to explain."

The kids are given meal vouchers to Quinnipiac's cafeteria, and so after the day is wrapped, William, Connor, Rahul, Ethan, and a few others walk to the dining hall to refuel. The mood is relaxed, jovial, the kids relieved that it's all over. There's some talk of how their presentations went, but that quickly dissipates.

Tomorrow, all the finalists will be back here for the nail-biting awards ceremony, a lengthy affair. But for right now, the kids are happy to eat some hot chili on a cold day and revert back to their normal teenage selves.

On Saturday, March 18, all the CSEF finalists, families, and teachers take their seats in one of Quinnipiac's gymnasiums. Sandy is at the podium and says, "This was a fair to be remembered." One hundred sixteen schools, 632 kids, 320 judges, and 84 volunteers were on-site this year.

The first awards announced are the specialty awards, those given and funded by corporations and benefactors. These prizes can be modest: Amazon gift cards, museum entrances, a few hundred bucks. But Andy's take is that it's all about getting to the stage, no matter the size of the prize. He can't sit still. He's got his camera out and he paces the gymnasium floor. Eleven years of these awards and he's no less anxious.

Early on in the specialty awards, Ethan starts cleaning up. He wins first place and $500 in the category of physical sciences. He wins the United Technologies Corporation award of $500 in UTC common stock. And he wins one of the awards given by the Connecticut Association of Conservation and Inland Wetlands Commissions.

The ATOMIC math awards are called.

"In second place, Steven Ma, Greenwich. In first place, Shobhita Sundaram, Greenwich." Shobhi also collects $300 as the second-place winner of the Xerox award.

After a long roll call of the specialty awards, the major CSEF fair awards are announced—the ones that determine who goes to Intel.

The top two winners in the categories of life sciences and physical sciences get Intel spots. Biotechnology gives one, as does engineering. And there's one spot given to a student from an urban school.

Biotechnology, always one of the toughest categories, is first.

"Fourth place, Sophia Chow, Greenwich," says the announcer. Gregg and Sophia are sitting in the audience together. It's unclear whether she's pleased or disappointed, as her face is steady throughout. She rises and walks to the stage to get her award before heading to the photo station.

It's an honor to get anything, for sure, but painful to be so close to the big award and miss. Noni Lopez, Andy's student who developed a diagnostic for Chagas disease, takes third place, and Michelle Xiong, who developed a novel, solar-based water filtration device, wins the category. The only place to go to a non-Greenwich kid is second.

Next up is engineering. Connor Li (already headed to Intel ISEF from his CT STEM win) gets fourth place and Bennett Hawley, a sophomore, wins third.

In the physical sciences category, Michelle Xiong wins fifth place, Agustina Stefani wins third place, and—no surprise—Ethan wins first.

As Andy's listening to the results, it sinks in. His kids have never won quite this big before. Of the seven Intel spots given by CSEF, four are going to Greenwich students. Ethan, Michelle Xiong, Rahul Subramaniam, and Luca Barcelo are all headed to Los Angeles for Intel. And that's on top of the three berths that Shobhi, Connor, and Agustina already won at CT STEM. Altogether, seven of Andy's kids will make the trip.

After all the photos and hugs (both victorious and sympathetic), Andy's eager to leave. He actually has to attend an evening lecture as part of the coursework he's doing for his administrator's certification, the 092.

Throughout this year, he's increasingly become interested in advancing his career. But he's torn. He loves the kids, both his research and ESL kids, but he'd like both more money and broader responsibility. He feels his dues were paid long ago. He's proven himself and now he wants to see what else he might be able to do.

A few weeks after CSEF, Andy's mentor and professor for his 092, Thomas Forget, Ph.D., decides he'd like to pay Andy's lab a visit. Forget (pronounced "For-zhay") is the former provost and vice president for academic affairs at Sacred Heart University and holds a doctorate in education. He wants to both see Andy's class and attend a meeting with Andy and his GHS mentor, the one who oversees his administrator's work and hours required for the 092.

Andy's been candid with Forget about his angst regarding his future. He shared with him what he perceived as the salary raise snub, as well as other things over the years that he feels have shown that the administration isn't as appreciative of him as he'd hoped.

One of the challenges is the sheer size of the school: nearly 2,700 students and some 400 faculty and staff. This means the faculty tend to be dealt with similarly to the kids: the highest achieving and most recalcitrant get noticed by the administration. So while Andy is better known than many teachers (thanks in part to the copious local media attention he and his kids receive), it's not as if anyone's lying awake at night mulling his long-term happiness.

What he's come around to is that he'd like to be promoted to oversee a districtwide science research program, to build out what he's created with the class. He'd coach and mentor science teachers in the Greenwich middle schools, help them get set up, and remain teaching his own research class. Everything he does would start earlier, hopefully sparking some passion while developing the kids' skills at an earlier age.

Forget's a savvy, experienced man who likes to cut through the bureaucratic cobwebs that can gather so readily in education. Andy tends to sit back on his good-guy heels and isn't forthright about advocating for himself. He's naïvely assumed that the commensurate attention and advancement opportunities would come to him—as if the system were an actual meritocracy.

Forget, at once warm and sage, feels no compulsion to be subtle or indirect.

So when he and Andy sit down with Andy's in-house mentor, Forget says that Andy is in a "unique and valued position in the high school. What do you think the district can do with and for Andy and his aspiration for school leadership?"

Forget later recounts, "Her head went down and she just shook it sadly and said, 'I don't think anything' . . . the structure was not flexible enough to find a way."

Forget asks again, "What do you think are the possibilities to get Andy a position of leadership?"

Her response: Then who will teach the honors [science research] program?

A former school principal, Forget throws out some possibilities, trying to pry open the thinking about how the district might be more nimble, make Andy happy—and retain him.

The mentor says, "Essentially, that would be impossible in Greenwich because of the bureaucracy, the newness of the superintendent." (There's currently an interim superintendent, but a new permanent one is starting in the fall.)

"My own read," says Forget, "is any new leader in an organization tends to be both autocratic and policy driven. Because that's a safe place to be. You don't exercise any degree of freedom until you have a level of confidence. That doesn't come in your first year, especially when you've got the kind of critical community that in Greenwich is watching every move.

"She recognized the play and acknowledged it with regret," says Forget.

But he thinks it's productive for Andy to hear for himself what his advancement prospects are (or are not) in Greenwich.

Forget isn't merely backing Andy as a political favor, as a hand trying to oil the rusty levers of bureaucracy. He sees what Andy does as a crucial lifeline in an education system that in many areas and many ways is crumbling.

"We've moved from education, teaching people how to think, to training, teaching people how to bark on time. And highly structured curriculum and even scripted curriculum in some places—the teacher reads the lesson. Those are not places where someone is being educated. It can't be.

"Look at the outcomes. The value is that people can think when they are freed to think, even seventeen-year-olds. Andy does not have a prerequisite requirement—you don't have to score a certain number on a standardized science test to get into his course. You have to have an interesting problem and some thoughts about how you're going to respond to that.

"Which is more valuable to the person and to the society? I can memorize something and give it back to you in an orderly fashion, even in a comprehensively well-expressed fashion. Or I can think. To me, it's not even a call."

44

William

So after his manic, twenty-fifth-hour hustle to get to CSEF, where was William? Oh, he was there, all right. In true William fashion, he was late, frazzled, and rushing to make it all come together, down to the hair of his Yinny Yin Yin.

The morning of CSEF, William is late due to a crash on the highway. Fortunately, Mark drove him here, a wise move given William's sleep deprivation.

"I went straight home [from the lab] and started and finished my poster," he says. "I didn't stop working for eight or ten hours. I woke up in the morning and I shipped it to the poster shop and finished working on my report. I pulled together my report and then I was looking for binders [for my paper]. I then rushed over to Quinnipiac to set up my poster."

In addition to the copious amounts of text the poster required, he had to build graphics to illustrate his work. Most of the other kids need quite a bit of help with their graphics, and Andy pitches in on this front. They produce and organize their data and then, in consultation with Mr. B., they decide what visual representation will work best. Andy often then builds the graphics using specialized programs.

William has once again boxed himself out of such assistance. But he manages to create dynamic, visually appealing graphics on his own. "Those are all him," says Andy.

Standing in the gymnasium now, William is visibly worn down. He's been fighting a bad cold for the past week. That, in conjunction with his manic sprint to the finish, has left him somewhat broken. He's in need of a haircut and so his hair's hanging limply over his ears. At least his outfit is smart and wrinkle-free. He's wearing a navy blue blazer, burgundy tie, white button-down shirt, and khakis. Predictable, but safe.

A pack of three judges approach him, two men and a woman. William perks up and launches into his presentation explaining the design, mechanism, and goals of his Alzheimer's diagnostic.

What ensues between him and the judges is a kind of motion-sickness-inducing drive through a winding canyon. One of the male judges says absolutely nothing. The other male judge and the female one have a lot to say.

Man Judge starts barraging William with questions, which quickly morph into his own lectures. Woman Judge tells him, "Zip it! Let the kid talk."

Man Judge ignores her. "Why a dipstick test? Why not a mega machine, a million-dollar machine?" This feels like a taunt of some sort, though it's unclear what the judge is going for.

Woman Judge says, laughing in a kind of crazy way, "Or genetics, or something? Check out the DNA."

"Make it expensive and complicated," says Man Judge. "Healthcare will like it."

Woman Judge's face grows sour. "So there's no cure for Alzheimer's disease, right? So I don't know if I would want to know if I'd be predisposed to it if we can't—"

William's looking for an in to speak. "That's a good question," he says.

To which Woman Judge retorts, "It's not a question. It's just a statement." She pauses and adds, "Even after I get it, I don't want to know I have it. So, I mean, what would be the value of it then, William?"

Ouch.

"Well, there currently are a number of therapeutic options available. Many of the tests for Alzheimer's treatment have failed primarily because they've been administered on patients who already demonstrate external symptoms of the disease," says William, trying his best to smooth things over without seeming desperate.

"Now, if we were instead to diagnose patients two decades prior to these external symptoms, prior to any irreparable brain damage, then there's a lot more facility for treatment and these same treatment techniques, the same therapeutic techniques. You know, this would also allow for patients to be determined who are predisposed and then allow the monitoring of the progression of this disease to allow for the future development of curative treatment," he says.

"So, we could all become guinea pigs is what you're saying?" says Woman Judge. Zing! She's middle-aged and has the sass of a diner waitress. Her comprehension of the actual meat of William's project is unclear.

"Well, if you were to put it that way," says William. "That's a smaller factor."

"They'd be paid guinea pigs," offers Man Judge.

Woman Judge tries to get serious. "Lemme just ask something else, William. I know we're being funny with you, but we gotta lighten up the load here.

"So, William, you mentioned that if you could detect it two decades ahead of it happening, so I'm assuming that over time the detection gets higher and higher, as you get closer to obtaining—"

"Yes, yes, but either way, this is designed to be a binary reaction," William replies.

"Okay, so this is yes or no, almost like a pregnancy test?" says

Woman Judge. Ding! Ding! Ding! Now we're cooking with fire. Something has registered.

"Exactly," says William, seeming relieved.

Not so fast, says Man Judge. "But in the case of pregnancy, it is yes or no. You can't be thirty percent pregnant. Right? In the case of Alzheimer's disease, it's more about hey, are you within the next two years or are you looking at twenty years?"

Woman Judge hops on this wagon. "Yeah, so how would you think about that, if you're doing a yes or no, that's where I was going . . ." She was? She is now that Man Judge has weighed in.

"I would actually be able to design a test, it would obviously be not as low-cost as this . . . ," William says.

Woman Judge pounces. Here it is! She's found the wonder boy's weakness.

"Ah, you see that!" she says.

"Now we're getting onto the same platform here," says Man Judge.

William keeps treading water, struggling to keep his head from slipping beneath the surface.

"What I would be able to develop is sort of an electronic system that would be able to measure the *concentration* of these gold particles," he says.

"Right, right, see that's where I would get to," says Woman Judge, as if she's now off to the races with William's technology, already herself developing version 2.0.

William says, "The more there are, you know, the more progressed."

"That's where I would get to," she repeats.

Man Judge says, "And I come from the engineering perspective and I look at things like probability of prediction." He talks about false positives, that is, tests incorrectly giving a positive diagnosis.

"You've got to deal with that position, as well," he points out to William.

"And that would be tested," William inserts.

Man Judge points out the potential downside of the false positives, saying that if you were to tell a healthy person he's got Alzheimer's, "you just killed him right there, anyways. You don't want that. And probability of prediction will change. Once you run your trials, you will be able to see it."

Man Judge adds, "But I think, to your point, there are certain dietary things which can delay the onset of Alzheimer's."

Woman Judge says, "Wouldn't it be nice though if they could tell you all those sorts of things ahead of time and you just did it that way?"

Uh, that's the WHOLE POINT of William's project. This is starting to feel like that Gravitron carnival ride, where a circular rotunda spins at great speeds, the floor bottoms out, and you stick to the wall.

Man Judge seems to be coming around, though not due to anything William's said. "It's good if you can detect it. There might be ways, hey, if it is really at that early stage, you can actually cure it," says Man Judge.

"Exactly. That's what I'm trying to, uh—" says William.

Man Judge interrupts. "So if you can get much, much sooner—"

Woman Judge interjects, "You're trying to say that, but he won't let you have a word in edgewise, right?"

Pot. Kettle. Black.

"William, it was really nice. I'm glad for you putting up with us," she says, weirdly. "I'm sorry." You get the feeling she's going to exit the gymnasium and head to the first happy hour she can find. All these smart kids really take it out of you.

Man Judge cannot turn off his need to expound. "What questions do you have for me?"

"Well, uh . . . ," says William.

"What questions do you have for me?" he repeats. "How can I help you?"

"I guess I'm interested in getting to know what your specialty is, your expertise?" says William, admirably reaching.

"My specialty right now is innovation coaching. I work with business CTOs and CEOs . . . ," says Man Judge, launching into his eleva-

tor description of his work. William does his best to look keenly
interested, though why a high school kid at a science fair would give
a rat's ass about innovation coaching is one of those mysteries
wrapped in an enigma.

All the while, Man Judge Number Two hasn't said a word. When
the odd trio of judges finally walks away, another kid comes up to
William, extends his hand, and says, "Let me be the first to congratulate you for winning CSEF this year." He's not being sarcastic. Anyone
witnessing this bizarre judging session would have to hand it to William for remaining calm and poised. It was exhausting to watch, the
kind of thing that makes your facial muscles twitch from the discom-
fort.

Outside the science fair world, one of the world's leading experts
on Alzheimer's disease had a different perspective on William's project. Dr. Murali Doraiswamy is a Duke psychiatrist and neuroscientist
who studies the biochemistry of Alzheimer's and closely monitors
diagnostic and drug development for the disease. He's the author of a
book for patients called *The Alzheimer's Action Plan*. After reading William's paper on his project, Dr. Doraiswamy wrote in an email:

> Conceptually, William Yin's project is brilliant—the sort of idea for
> which the NIH or a VC firm may have given a $3M seed grant. Of
> course, his project just demonstrates the in-vitro (test tube) feasibility
> of this—not in actual humans. We don't know the relationship be-
> tween brain soluble amyloid and urine amyloid, and it will take 2–3
> years of research to even determine if this is feasible in humans. How-
> ever, if it turns out to work, it would be the low cost, scalable, early
> detection test of brain amyloid pathology that the field is so desper-
> ately seeking. It is mind blowing that a 17-year-old could conceive of
> this—and kudos also to his mentors.

Even before his disastrous encounter with the judges, William has
been torn about whether to attend the awards ceremony, which hap-

pens to coincide with a band concert. The band director, the polarizing Mr. Yoon, is infamously uncharitable when it comes to kids missing performances, which puts those who do many activities in the tough position of feeling like they're inevitably letting someone down. William is well aware that Andy thinks it looks very poor when your name is called to accept a big award and you're nowhere to be found.

So in an effort to try to please everyone, William decides that a possible solution is to arrive to the awards ceremony early, gingerly approach Sandy, explain his predicament, and poke around the edges to see if she'll slyly indicate to him where he stands. He decides that if he's going to get a big award, he'll stay for the ceremony and scratch the band concert. If there's no glory to be had, he'll duck out and go play his mellophone.

He goes up to Sandy and says, "I have a conflict—"

"You're not going to ISEF," she tells him, straight up.

This is a tough pill to swallow and it doesn't go down easily. It's not helped by Sandy's delivery, which William refers to as a "blunt statement." Arguably, though, he's lucky she told him at all ahead of the official ceremony. (He winds up winning fifth place in two categories, engineering and biotech. Respectable, but not high enough to get to Intel.)

His mind pools the information and its ripple effects. He did all that work only to place the lowest he's ever placed at CSEF. There will be no Intel ISEF, no addendum to his Harvard application. How great would it have been to notify Harvard that his astonishing Alzheimer's diagnostic was sending him to Intel ISEF for the third year in a row?

In the past, all the late nights and boatloads of stress paid off at this very moment. He backed himself into a corner, playing it dangerously close, only to be propelled forward, rewarded, even, in glorious Yin fashion.

But today's not William's day.

SPRING

45

- - - - - - - - - - - - - - - -

Romano

Even before Andy gets to CSEF, he's thinking about what comes after. Or rather, who.

"I really want to spend some time with Romano," he says. "He worked really hard, it didn't pan out, but I'd love to get something going for him. He's such a great kid."

So once CSEF is behind him, Andy pulls Romano aside and says, "Let's talk."

The two of them are sitting at one of the low tables. Andy clasps his hands and takes things back to the drawing board. "So you want a liquid bandage? Here's what you're going to do. Go to CVS and buy some liquid bandages. Let's take a look at these things," he says.

Romano's so excited, if he could go to CVS via jet pack, he would. Even though he practically had Andy at hello, he badly wants to show his teacher he's worthy of being in the class, that he's good for more than bringing in tasty Italian food and elevating the social atmosphere of the lab.

The next day, the two of them are looking at a few samples of liquid bandages. One of them is an aerosol spray.

"It's extremely volatile and smells like rum," says Romano. "It contains nitrocellulose, nasty shit." Nitrocellulose is highly flammable. "I don't like the toxicity, and it provides no healing benefits."

Another aspect Romano doesn't like is that when sprayed onto the skin, it forms a very thin, skin-like layer. It's not at all the sturdy, firm, gel-like bandage Romano has envisioned from day one.

The other brand he and Andy are looking at requires a little brush for application.

"This doesn't stabilize the wound. If I'm shot in Afghanistan and I have to take out my little brush . . . ," says Romano.

He decides to run the contents of these bandages through the FTIR to get a chemical footprint.

As suspected, neither of these products contains antibiotics or any kind of healing ingredients. Romano still thinks this is where he can introduce something novel.

He focuses on the consistency issue; he still wants to produce something thicker in substance. He and Andy circle back to Andy's suggestion months earlier to try using HydroMed, a gel in which the liquid component is water. HydroMeds come in various forms. Andy happens to have clumps of the stuff in the form of squishy beads, leftovers from previous projects.

Romano takes a small clump of beads and plops them into a beaker filled with water.

The next day, as suspected, some of the clump has dissolved into the water, making it a bit milky in color. Romano puts some of the water on his skin and within eight minutes he's psyched when he sees that it dries and hardens. However, the covering is thin because it seems only trace amounts of the HydroMed actually dissolved into the water. So consistency-wise, it's similar to the commercial liquid bandages he bought.

Just to be sure, he runs the water through the FTIR, and indeed, the HydroMed shows up, but not in huge amounts.

"We have something that's thin and doesn't have any healing benefits and takes two minutes to cure. And it's not convenient," says

Romano. But unlike his many L-DOPA misses, Romano's beginning to think he can make something with this HydroMed. The fact that it even reacts with water seems promising to him. Even though it's not the ideal consistency, he wants to see what, if anything, it can do on the healing front.

"Let's get to work," says Andy.

"Let's do this," says Romano. "Let's get an antibacterial in here."

Andy has some left over tetracycline, a common antibiotic, in powder form.

Romano has a fresh beaker filled with water and he pours tetracycline powder into it. The water turns Gatorade yellow. He takes an FTIR of the water and then drops in a clump of the spongy HydroMed beads to see if they will absorb the antibiotic.

To assess this, Romano runs water samples through the lab's UV/VIS spectrometer, which uses color and light to measure chemical contents. Every fifteen minutes, Romano takes a sample of the water surrounding the HydroMed that was once saturated with tetracycline and runs it through the spectrometer.

He's blown away by the capacity of these instruments to work for him.

"If something's really yellow or colorful in general, it's going to absorb more light than if it's clear," says Romano.

The absorption goes from 1 to 0.4, showing that the tetracycline is leaving the water and getting absorbed into the HydroMed. As time passes, the water becomes less and less yellow.

One snag is that the tetracycline in water is so yellow, Romano has to dilute it hundredfold. If he doesn't, the spectrometer won't be able to read the samples. So he carefully dilutes all his samples and adjusts his calculations accordingly.

He's doing all this both in class and after school, working with an intense focus. This momentum is what he's been wanting all year and he doesn't want anything to bottom out again. And he knows he has a way to go.

"At this point, I've told you how much tetracycline was absorbed by the HydroMed," he says. "This is not even step one, it's like step zero point five."

Outside the lab, he's working on other crucial matters. The junior prom is April 22. The venue where it was meant to be held had a fire a few weeks ago, so the prom will now be held in the GHS student center, i.e., the cafeteria.

Romano has waited pretty late in the game to even think about a prom date. He's been hanging out a bit—not in an overt romantic way—with a half-Chinese, half-Caucasian basketball player we'll call Katie. He likes Katie—she's cool. Not a mushy type, no fawning, a jock. She also happens to be one of Sophia Chow's best friends. Romano's thinking she might be the prom date for him. So, one day as they're walking up the stairs in the student center to Clark House, he asks Katie, "Do you have a date for the prom?"

"No, do you?"

"Would you want to go to the prom?"

She says yes, to which he says, "Cool."

Done. Signed, sealed, delivered.

But Romano's old-school and not about to let that be it. Of course, he's going to prompose—and he wants it to be original. He enlists a crew of his boys. He hits up the basketball coach for a favor. And he gets Sophia to pitch in.

On Thursday, April 14, at around nine P.M., Romano and three of his boys roll up to Katie's house. It's dark. They keep the headlights of the car on. They get out and start blasting "Get'cha Head in the Game," a song from the movie *High School Musical*. Romano's wearing Katie's basketball jersey and his backup dancers all have GHS b-ball jerseys, too. Sophia videos the action.

Romano and the guys dribble basketballs and start doing a choreographed routine, which they've rehearsed over the past couple of hours. This song is performed on a basketball court in the movie, too, but Romano's choreography is original.

Katie is sleeping, likely bushed after a tough week of practice, but hears the commotion in her driveway and walks out of her house. Sophia tells her to come over.

"Are you serious right now?" Katie asks. Looking closer, she says, "Is that my uniform?"

When Romano sees her, he bounces over to her and says, "What's going on? What's going on?" and then goes back to the routine.

The fast-paced song slows to ballad tempo with the following lyrics:

> Why am I feeling so wrong
> My head's in the game
> But my heart's in the song
> She makes this feel so right

For this, Romano sings along, strolls over to Katie, and says, "Will you go to prom with me?"

She says, now fully awake, "Yes." The gang cheers and claps. She and Romano hug. It's all very meta, since this whole scene could absolutely be part of its own movie.

There's a GHS junior class site where people post their promposals, which are almost exclusively still photos. Over the next few days, Romano's promposal video goes viral, and the consensus is that he crushed it—he went all in and it totally paid off.

His kudos continue in the lab, where people, including William, give him compliments, telling him his promposal was unreal.

With this bit of business taken care of, he's not letting up with his bandage pursuit. Now that he's established the HydroMed can absorb the antibiotic tetracycline, he wants to see if it'll also release it.

"If it releases none, the wound gets no benefit," he points out.

The clump of HydroMed is now bright yellow because it's saturated with the tetracycline. Romano drops the clump into a ten-milliliter test tube. He places the test tube in a warm water bath to mimic skin temperature.

Again, in fifteen-minute increments, he uses a pipette to remove small amounts of water to run through the spectrometer. And sure enough, over time, more and more of the tetracycline is being released into the water.

As he's running his samples, and each time seeing the antibiotics increasingly leaving the clump and being released into the water, he's got a triumphant grin on his face.

This payoff, this satisfaction, the feeling that he's getting something for his work, brings him unabashed joy. When science goes your way, he's learning, you're in a whole new world. And this is the world he wanted to be in. When his L-DOPA project was nothing but a series of fails, he couldn't help but look around the room and compare himself to others in the class.

"It felt a little demeaning," he says. "All these kids are curing cancer and I'm wrapping sausage casing around a test tube.

"Now I have HydroMed that can be loaded with an antibiotic *and* be released, thus providing healing benefits to the wound," he says.

But he's not finished yet. While the basic mechanism appears to be viable, it's not ready for practical use. No one's going to pour liquid the consistency of water onto their skin—that was Romano's beef with the ones already on the market.

No, he still wants that gel.

The prom, as you might know, is really not about the prom itself. It's all about the pre-prom and the after-parties. The pre-prom is all about the photos, when everyone's glammed up, the mascara is fresh, the hair has no flyaways, the tuxes are crisp, and the shoes are super shiny. The after-party is all about where you can drink the most, do the least damage, and not get busted by the Greenwich cops, who are on high alert every prom night. Not everyone engages in drunken debauchery. Andy likes to point out that one year, four research kids, not having an after-prom plan or party to go to, went back to one kid's house and watched *Sharknado*.

There's some stickiness that arises in regard to logistics. Katie, you see, is not in Romano's friend group; they run in different circles. There's a lot of debate about whose group they should spend their pre- and post- time with. As per usual, the discussion takes place between Katie's pack of girlfriends, who band together to tell Romano, "It's Katie's night." But he feels it's his night, too. So they cut a deal to split their time between both pre-prom parties. As for the after-party, Katie just wants to go hang with her friends. Romano will go to an after-party with his friends and then head over to hers. It's all about the diplomacy.

He decides that before he and Katie head to Burning Tree Country Club, site of one of the pre-parties, he wants to do pictures at Nanni's. At the Orlandos', Rome helps his youngest son don his tuxedo jacket. Romano has chosen a white bow tie, white vest, and black jacket.

The two Romanos stand in the foyer and look at the younger one in the mirror.

"Look at the shoes shined. You look good, son," says Rome. "You dress up nice, cleaned up nice."

"What do you think, Mom?" Romano calls out to Kim, who's in the kitchen.

She sees him and says, "Oh, my god." She's on the brink of tears but holds it together. They share a tight hug. She cried seeing the promposal video. Lately, all of these rites of passage are compounding and pointing to the inevitable: her baby boy is growing up.

After Romano and Katie take pictures at Nanni's, where a massive Italian spread of homemade cured meats, cheeses, olives, and pastries has been set out, Romano and Katie depart for their first pre-prom appointment at the Chows'.

Kim and Rome set off in Rome's black Lexus, with its vanity license plates that say simply "Rome." They snake their way to Greenwich's far mid-country, the cutoff between mid-country and backcountry being the Merritt Parkway, which runs east to west through the state of Connecticut. Burning Tree actually sits against the Merritt, so for all that money, you can golf, in part, behind a highway.

The country club is filled with parents, friends, siblings of the promgoers. It's been a rainy, overcast day, but it's not raining at the moment. Groups float between the back patio and an indoor room where people gather around high-top tables. Ooohs, aaahs, and compliments abound. And of course, there's a barrage of photo taking.

Eventually the whole lot is herded out back for big group photos: girls only, boys only, and then everyone with his or her date.

Romano and Katie are the last to turn up and just miss the big group pictures.

Romano rushes out back, and the first person he sees out on the patio is Matt Armstrong, his longtime friend. You may recall that, back in the fall, Matt and Romano got the party started at the homecoming dance. Matt's date, like Matt, has Down syndrome. She wears a pink shawl over her shoulders, a pink dress with a sheer overlay, and gold lamé sandals. Her hair has been professionally styled.

"What's up, Matt?" says Romano, moving in for the bro hug. Matt introduces his date, and Romano tells him how pretty she looks.

"Yes, she does," says Matt, all pride.

The kids make their way to a rented school bus procured by Kim. The mood on the ride to the high school is one of nervous anticipation and fun, and free of any drunken antics because no one's consumed a drop of alcohol for one simple reason: every single kid gets Breathalyzed at the door. The first thing the kids do upon arrival is open their mouths and breathe into the device, which resembles a flashlight. If it blinks green, they're good to enter the prom.

Romano's posse doesn't want to stay that long, and it was a big negotiation to decide what time the rented school bus will set off for the after-party. The kids want to mosey as soon as possible so they can drink, dance, and play beer pong. As for the family hosting the after-party, they've had all the kids invited sign a contract (that, of course, has no legal bearing) stating a long list of rules and regulations. Every infraction is exactly what you sign up for if you throw an after-prom rager at your house. The kids all say there's some weird legal loophole whereby if the parents don't provide the booze or actually witness the kids imbibing it, they're not liable.

So it's up to the kids to obtain the alcohol (not that hard between older siblings and a well-known list of local liquor stores that will sell to anyone) and do things like stash it in the bushes and backyard ahead of time. The parents are home but willfully remain away from the action.

Romano likes the occasional buzz but never wants to get obliterated. It actually bums him out when he sees his friends drinking themselves to blackout levels.

"Why would you want to do that?" he says. "It's just not fun."

He splits his after-prom time between his friends' party and Katie's. But it's so late when he gets to her party, she's tired and only stays for about an hour more. There had been the possibility of a kiss, but the momentum has fizzled from a combination of going their separate ways and the late hour.

Still, when it's all said and done, he feels he had a good time. Was it the best night of his life? No, not even close, especially after all the buildup. But, yeah, it was good.

Romano's back in the lab on Monday, still puzzling about how to change the consistency of his too-liquid bandage material. He's shown that the HydroMed absorbs the antibiotic and when placed in water, it releases it—which it would need to do on a wound. But it's still just too watery for practical use.

He explains his dilemma to Andy.

"How can I make it thicker? How can I get more HydroMed to dissolve into the water?" he says.

Andy suggests he try putting his clump of HydroMed loaded with tetracycline into liquid acetone, which evaporates much more quickly than water. This makes sense to Romano, the AP Chemistry whiz.

Romano does this, but the HydroMed-acetone concoction is still really thin. He tries putting as much HydroMed as possible into an acetone-filled test tube and shaking it rapidly.

"I'm trying to get the HydroMed to break apart into the acetone to make it thicker," he says.

When this fails, he decides to try something Andy mentioned as an option: sonication. A sonicator uses sound waves to break up molecules. There are a couple of different designs, but the one in the lab has a small vat that holds a water bath.

Romano puts his HydroMed-acetone mixture into a small beaker, closes the lid, and turns on the sonicator.

He takes a seat at one of the lab's low tables, fishes his homework out of his backpack, and starts working.

Lou, one of the custodians, walks in and he and Romano shoot the shit. Romano has befriended all the custodians and security guards, knows everyone on a first-name basis.

After an hour, he gets up to check on things. He opens the vat and peers down into the beaker. The clump of HydroMed is completely gone.

"It looks like Purell," says Romano. He gets a stirrer and pokes it.

"It's like a gel!" he says. He immediately calls out for Andy, who makes his way over.

"Look what just happened!" says Romano, who has removed the beaker and is showing Andy his thickened yellow mixture, which has the consistency of gel toothpaste. It's actually thicker than Purell. Andy's super gratified. This is one of those times when he himself didn't know whether an idea would work, and it's so sweet when it does.

Romano smears some on his skin and within seconds, it's cured.

"I would totally buy this, this is great!" he exclaims.

He runs out of the lab, into the hallway, and does a victory jump, fist high in the air. Later he says, "I was imagining a camera pausing as I got to the peak of my jump and that ended the movie."

This means Romano's going to make it to a science fair, after all. Following his many L-DOPA fails, the pigskin, his completely insanity-producing video bungles, working with the smelly gelatin—his day is finally going to come.

The Norwalk Science Fair is Friday, April 28, so he's got a few days to write up his poster, get it printed, and rehearse. He's beyond himself with excitement, but also a complete heap of nerves. He gets a little twitchy thinking about presenting at a science fair. Sure, he's acted onstage, he's constantly putting himself out there to practice his Italian, but Romano has a certain reverence for the science research kids and it's hard for him to place himself in their company. Now it's all staring him in the face.

The fair is held at Norwalk Community College. It's like the fast casual of science fairs. The kids are set up at numbered stations in actual science labs. There's no big gymnasium or hall.

Rome drives Romano to the fair in the family's enormous SUV, which is the only vehicle that can transport the four-foot-high poster mounted on foam. Romano spent the night before rehearsing his presentation in front of Rome, refining and smoothing his words.

The morning of the competition, he turns up in a blue suit and mint-green tie. He's so nervous, he plunks down his stuff at the first table he sees. He arranges everything with precision, repeatedly moving things to see where they look best. A girl approaches him.

"I'm number one," she tells him.

"I'm sorry, what?" Romano says.

"I'm assigned to number one," she says. "You're in my spot." She points to the piece of paper with the giant "#1" printed on it.

"Oh, there are assigned spots? Okay, okay," says Romano, who quickly starts gathering his things to find his own. Amateur move.

His actual spot, number seventeen, is near some fellow Greenwich kids. He neatly arranges his poster, notebook, abstract, and glass slides with his hardened gel. He paces the floor, practicing his presentation. He chats with Charlotte Hallisey, who's in his row.

The wait feels interminable. But finally, three judges approach and say, "Romano Orlando?"

He launches into his speech, explaining the need and benefits of a liquid bandage. He talks about how he introduced the antibiotics into the HydroMed and then was able to prompt their release. He makes use of his poster, pointing out graphs that show his experimentation metrics.

He's talking a little fast, but the presentation is smooth. The judges nod with approval, seeming impressed.

When it's time for questions, one of the judges quizzes Romano, asking, "What does FTIR stand for?"

"FTIR stands for Fourier-transform infrared spectroscopy," he replies. Phew, he thinks, grateful he thought to memorize that. So clutch.

Elsewhere, the kids are scattered throughout other lab rooms. Andy has made sure everyone's comfortable and feels good. Even though this is a much smaller, less fancy fair, with fewer than one hundred entries, he approaches it with the same enthusiasm as he does the big ones. He's also amped up because today represents his very last push with projects. Everything and everyone's done now.

"I feel good, real good," he says. "This is it, man. This is it. I'm not looking at another poster."

When it's time for the awards, the fair director steps up to the head of the room. There are no keynote speeches, no endless thanking of sponsors. It's utterly refreshing.

"We will be awarding eight honorable mentions and then three top prizes," says the fair director. Moments later, he says, "Third honorable mention goes to . . . Joe Konno." Joe! Joe has worked all year in Andy's lab on a complex hepatitis C project. He's the shy, bespectacled kid with braces who's also a champion violinist. He sits in the lab with Romano and Collin.

"Sixth place honorable mention goes to . . . Romano Orlando." Romano stands up, straightens his tie, and walks to the head of the room to collect his certificate and prize. He may as well be accepting the Nobel, he's so puffed up and pleased.

When the top three prizes are announced, it's a clean Greenwich sweep. Michelle Woo gets third place for her project that converts CO_2 to biofuel, Alex Kosyakov wins second for his battery, and the top prize goes to none other than Collin Marino. Applause and back-slaps all around.

Andy has a look of utter satisfaction and fatigue on his face, like he's just finished Thanksgiving dinner and is about to plop down on the couch for hours of football viewing. Mindful of those who didn't get awards, he never jumps up and down, but inside, he's very happy.

So is Romano. "Oh, my god. So happy, so happy," he says, grinning. He later says that placing in the top ten was, by far, one of the best moments of his high school career—based on the sheer effort involved alone.

The kids head upstairs to the labs to collect their things. By the time Andy steps out into the spring sunshine, Romano's sitting on a ledge, legs dangling, waiting for Rome to pick him up.

"Thank you, Mr. B. Thank you so much," he says.

"What are you thanking me for?" says Andy.

"It was really great, Mr. B. Thanks for everything," says Romano.

"You have a ride? You good?" Andy asks.

"Yeah, yeah. I'm good. My dad's on his way."

"All right. See you Monday," says Andy.

"See you Monday."

It's a convertible kind of day, so as soon as Andy gets to his car, he puts the top down on his Volkswagen Eos. He gets in and looks over to Romano, who's all alone and seems to be somewhere deep in his own mind. He's got his head tilted toward the sun, his eyes closed, and a beatific smile that is utter contentedness.

He looks absolutely golden.

46

- -

William

March 3, 2017

Dear William Yin,

Congratulations! It gives us great pleasure to inform you that you have been admitted to Duke University and selected as a finalist for the Angier B. Duke Memorial Scholarship. Yours was among the strongest of more than 34,300 applications for admission to Duke and your official letter of admission should arrive at the end of March. We couldn't wait, though, to tell you about the A. B. Duke Program and give you the news that you're a finalist.

The A. B. Duke Scholarship, Duke's flagship merit award, provides 100% of tuition, mandatory fees, and room and board for four years (eight semesters), a six-week summer program at Oxford University, and access to A. B. Duke–specific research/special project funding. The program is designed to enable you to realize fully your intellectual ambitions, your individual talents, and your particular interests—from environmental studies to public policy, from philosophy to genetics, and so much more.

This is potentially a huge boon for the Yins. A great deal of William's decision on where he goes to college is dependent on financial aid and scholarships. The family owns a rental home in the less desirable western side of town, and Mark has said they could sell that property to fund William's higher education, though that doesn't appear to be an attractive option.

The roughly forty-five A. B. Duke Scholarship finalists will be flown down to the university's Durham, North Carolina, campus and hosted there from March 29 to April 1, where they will undergo rigorous interviews. Twelve finalists will be awarded the full-freight scholarship.

William will hear from the majority of other colleges he applied to on April 1, while he's at Duke. He's already been admitted to Stanford and Yale but hasn't learned what financial aid he may receive. It's become clear to him that now the focus is shifting—not on where he gets in, but how much money is attached to any offers. And at the end of the day, money will be a huge contributing factor.

So he knows that during the three-day Duke interview process, he has to swing not merely for the fences but for the next town over. As prestigious as it is to be a finalist—as Duke stated, he's at the very top of a heap of 34,300 kids—it's not enough. He needs to win. How fantastic would it be to be able to attend college knowing that when he's tossing his graduation cap in the air at commencement, he and his family will be debt-free?

When William arrives on campus, he knows a few of the other finalists. This is not surprising thanks to his frequent mingling with America's academic elite at the science fairs, his summer MIT program, music competitions, etc. He is the only A. B. Duke Scholarship finalist from Connecticut.

Duke has packed the kids' schedules, from morning until evening, for the three days. They stay with current scholars, and William is placed with a science kid, someone who applied early to Harvard, was deferred and then rejected, but was also accepted early to Stanford and Yale.

William is still holding out hope for Harvard, an excruciating wait that will come to an end in just three days. But he's trying to put this out of his mind. He needs to focus on why he's here.

The interview process, he says, "is really, really intense." To hear this from him suggests this is not something most people could endure. There are panel interviews, which he says are to "test how well the student thinks."

When he tells the panel he's interested in neuroscience, the discussion somehow turns to criminology and fascism before winding its way around to biochemistry and ethical issues in science.

"They pride themselves on the fact that people [on the panel] like to ask pressing questions," he says. It's a "lightning interview" format where the kids are given five minutes to digest the question and consider their answers. He's gotten plenty of practice for these kinds of grillings from his experience on the science fair circuit, so he's not easily thrown.

One question posed to William is, "If one has the ability to normalize beauty or intelligence, should one do it?" This is followed by a cascade of other inquiries—some of which he has to write essays for—that include what science research project is he most proud of, what course he'd want to teach at Duke, is there a question that drives his intellectual curiosity, etc.

One of the tougher essay questions, which is due at midnight on the day it is assigned, is a problem-solving one. The year is 2050 and earth is about to destroy itself. You're tasked with sending a force of one hundred individuals to Mars to establish a colony. How would you go about doing that, how would you establish a colony, and what remnants of life on earth would you keep?

"I unfortunately started that essay at eleven forty-five P.M.," William confesses, "so I just wrote as fast as possible."

He writes that he'd want to keep as much of society the same as possible, in order to retain a sense of civilization. "I sort of related it to *Heart of Darkness*. Humans try desperately to retain threads of what they're familiar with in order to maintain their sanity. Looking at the long-term stability, I wanted to establish as stable a colony as possible by retaining social and cultural norms and forms of government, a representative democracy."

When he and the other finalists compare notes, he thinks his answer was kind of bland. Some had a more sci-fi bent, mandating things like a male-female ratio of one to ninety-nine in order to insure genetic diversity. (One way to look at it.) Others took a rather draconian position of exterminating people who couldn't perform their jobs well. (Harsh.)

Duke does its best to fan its feathers, making sure the finalists know that its food is rated the best in the country, that it recently spent $95 million on its new dining facility.

"I had the best Indian food I've had in my life," says William, adding, with a note of intense gravity, "There's a dedicated crêpe bar."

William has his own agenda of things to look for while he's here. For one, he wants to make sure the A. B. Duke Scholars are a cool, interesting, diverse group. He doesn't want to feel like he's joining some self-isolating nerd cult. Because the scholars can basically design their own course of study, William wants to see if they still mingle with the mainstream Duke cohort. He sees they do, which is a relief.

He's also kind of bowled over by the sense of Blue Devil spirit and pride.

"You never see a guy going around saying, 'I'm a Harvard student, go, Harvard.' You'd think going to an elitist institution, they'd be full of pride, but it's the opposite. Going to one of these elite institutions means people will segment you [off from the rest of the population], so you can't tell anyone. You'd say, 'Oh, I go to a school in Boston.'

"But it's not the case with Duke. Duke is very well known and their [social media] profiles show them shouting and cheering and lots of school spirit. I know I won't find that at an Ivy League school," he says.

He also thinks his fellow finalists seem collaborative and genuine. In other settings of the academic elite, he has found that "some kids are playing the networking game of being nice" and projecting what he calls "controlled charisma and pragmatic charisma." This group seems more human all around, which he finds very appealing.

One unfortunate aspect to the three-day audition is that Duke times it so that the finalists are there during regular admission notification from most schools. In theory, you could be enjoying a chocolate crêpe at the dedicated crêpe bar and learn your fate.

The decisions are slated to land at five P.M. on April 1. By five fifteen, the kids will be at a farewell gala, having just learned the decisions of the other schools to which they've applied.

William's in his host's dorm. He changed into his gala attire at five. He takes his laptop to the dorm hall kitchen and runs through the emails at rapid pace, not lingering on any one outcome. The whole way this is going down is very much in line with William. He likes to tackle big milestones alone and especially now, especially this, he gets to do whatever it is he might do and feel whatever it is he might feel in private. In advance of the notifications, William later says, "I think I forced myself to be completely emotionless. I suppressed all my emotions." He decides to check Harvard first, "to get it over with."

He opens the email. Rejected.

Next up is Columbia University. Wait-listed.

Northwestern. Rejected.

Carnegie Mellon. Rejected.

But after that, it's a string of acceptances, to MIT, Princeton, Dartmouth, Cornell, University of Pennsylvania, and Brown. He's also been accepted to all of the prestigious state universities to which he applied. (In an interesting twist, Derek Woo, the Regeneron finalist, is accepted to Harvard and Columbia but rejected from the six other Ivy League schools—the exact opposite of William. Derek opts for Harvard over MIT.)

He snaps shut his laptop and heads to the gala, where no one is able to be present in the moment, as everyone is still processing his or her acceptances and rejections.

When it's all said and done and the William Yin road show is over, he flies back home and practically stumbles through the front door, mentally and physically spent. Moments after he walks into his house, his phone rings.

It's Duke. They're calling to offer him one of the twelve A. B. Scholarships. He's floored. He did it.

He could go to college for *free*—an offer valued at $266,956.

And that doesn't even count the study abroad in England and money to do research.

But this is William. He doesn't jump to accept. No, he needs to think about it. He needs to see what kind of financial aid or scholarships other places might offer.

As he and his parents start wading through the offers, it turns out that no school even approaches Duke. Stanford offers what will amount to half of its $64,782 price tag for the 2017–18 year. The financial aid packages from MIT and the six Ivy League schools are so inadequate by comparison that William pretty much instantly scratches them off his list.

It's come down to Duke or Stanford—the guaranteed full ride or what looks to be a half ride (although it's unknown whether Stanford will offer William half the cost for the entire four years).

He knows Stanford, as a policy, doesn't match merit scholarships, but he decides he'll call the regional admissions office, mention his Duke offer, and see if they'll pay for him to visit again or increase the financial aid.

The first day he's back in the lab, AP Chemistry teacher Shirley Barban walks in and says to William, "Made a decision yet?"

Now that he's been rejected by Harvard, William has eased off his stark secrecy policy around his Stanford acceptance. For the rest of April, he's surrounded by a chorus of people asking what he plans to do.

At lunch one day in town, his father, Mark, seems relieved that the end of the college brouhaha is nigh. The family has suffered a bit of whiplash from it all.

Mark says the decision is William's and his son must go where he thinks he'll be happiest. His mantra to his kids has always been, "Your health and happiness are the two most important things."

There's often an inverted power dynamic in immigrant families where the American-born kids, more linguistically and culturally fluent than their parents, essentially call the shots. Some of this seems to be at play with the Yins. It's likely also fed by the fact that given William's performance in everything he does, it's hard as a parent to argue with his methods and choices.

Mark admits that Harvard mattered more to Wendy than anyone else in the family—and that she was upset by the rejection. Mark is mindful of the geographic distance that looms if William goes to Stanford. "From a parent's perspective, I would hope he stays close," says Mark. William's grandparents and family friends have given a thumbs-down to Stanford on the grounds of distance alone.

Mark imagines that if William goes to California, Verna will follow and they'll ultimately decide to stay on the West Coast. If that's the case, he says, he and Wendy would likely make the move themselves.

As for the money, Mark says, "I don't focus on the economics, I focus on what's best for his future." This doesn't appear to be a bluff. You get the feeling that Mark, who came to America needing to borrow three hundred dollars, would wash cars or collect garbage in order to give his children the best education possible. Whatever it takes.

As starry-eyed as William was about Duke, over the next few weeks, the luster fades. The crêpe bar wasn't enough. It seems every day he inches closer and closer to Stanford. People question whether he's being sucked in by the name and the perceived prestige.

Andy's the opposite of wishy-washy on the matter and more than willing to express his opinion.

"I put '300K' in big numbers on the board and I said that's the number of reasons to go [to Duke]," he says one day. William's retort was "If I take the Duke money, then my parents won't be on the hook for any tuition," which could adversely affect Verna's financial aid. Whereas, if his parents are having to pay half the cost of Stanford, that might increase aid to Verna.

"Here's the reason why that's bullshit," says Andy. "Three hundred thousand reasons. He won't decide until eleven fifty-nine on April thirtieth." He cautioned William not to get seduced by the prestige thing. "You're going to spend a hundred and fifty K on that name," he says, pointing out that Duke is not $150,000 worse of a school than Stanford.

William and Shobhi are headed to the JSHS nationals in San Diego on April 26. He decides to commit to a school before he departs. And so it is that on April 25, William tells Stanford that he will be joining the Class of 2021.

47

Sophia

Everything about Sophia on prom day is so characteristically her—elegant and understated. She didn't agonize over what to wear, choosing a $110 periwinkle blue lace cocktail dress at Bloomingdale's, which she likes for the price as much as its simplicity. She watched mildly horrified as some of friends dropped $400 and $600 on dresses they'll wear for a few hours before changing into their carefully curated after-prom outfits.

Sophia and many of her friends have gotten spray tans, which aren't exactly subtle. But everything about Sophia's look has an unfussiness about it. Save for a few select tendrils, her hair is mostly down and the edges have a light wave to them. Her makeup is barely there.

Taylor, her date, has gone for the black tuxedo and silk black straight tie rather than bow tie. They pose together for photos and then separate to go take pictures with the huge assortment of kids and their families. The boys are a kind of smug cool, while the girls are giddy, everyone squealing and showering everyone else with compliments galore.

"Oh, my god! I LOVE your hair!"

"That is SUCH a good color on you!"

"Oh, my god! You look AMAZING!"

But, alas, like Cinderella at the ball, Sophia's prom date, Taylor, basically evaporates after the prom. Somehow all the vibe was in the anticipation, but following the actual evening, they suddenly have nearly nothing left to say to each other. Sophia's actually fine with this and Taylor appears to be as well.

Besides, Sophia's focused on other things. On March 20, she was notified that her Lyme disease project was accepted to ISWEEEP—the International Sustainable World (Engineering Energy Environment) Project—a prestigious, rapidly growing global science fair, now in its tenth year. With 400 competitors to Intel ISEF's 1,700, it's a more boutique competition. It also has a much smaller prize purse, paying out a total of about $100,000, compared to ISEF's $4 million. But it's an international affair that draws kids from sixty countries, and each year, Andy says, it's getting more slick and well produced.

Held in Houston over four days, ISWEEEP requires kids to be chaperoned. Every year, this presents a conundrum for Andy. GHS won't pay for him to accompany the kids (which you could argue is weird, like sending the football team off to play without the coach), and he refuses to pay out of his own pocket. He also doesn't want to sap the research fund, which is earmarked for project expenses.

This year, six of the eleven GHS kids who applied have been accepted to ISWEEEP, the most he's ever had, another record in what's turned out to be Andy's banner year. Andy's hoping that some of their parents will offer to chaperone, relieving him of the hassle with the trip funding. One by one, the kids all come back and say their parents can't do it.

But Gregg Chow raises his hand and then so does Anthony Minichetti, father of Dante Grace. Gregg decides the right thing to do is to pick up the cost of Andy's plane ticket and then send a gentle solicitation email to the other parents of the ISWEEEP kids, asking if they'd like to contribute. (He gets one check from a parent.)

Andy has earned a lot of cred with the ISWEEEP folks over the years, as his kids have consistently won top prizes. This year, a fair administrator comps a room for "the great teacher," which makes Andy totally psyched. So between Gregg picking up the cost of the plane ticket and ISWEEEP giving him a free room, Andy's good to go.

Sophia had spent hours refining her application materials and poster, getting feedback from Andy, rearranging it, and polishing it just so. When she found out she was accepted to ISWEEEP, she couldn't believe it. She was finally going to get her turn on the bigger science fair circuit. Even though she's had a solid year here in Connecticut, placing at both CT STEM and CSEF, she's wanted the validation of a larger-scale science fair, with all the frills and glamour. She was in school when she got the email informing her she'd been accepted and immediately left her chemistry class to go tell Andy. She will be joined at ISWEEEP by Dante Grace Minichetti, Bennett Hawley, Sanju Sathish, Rahul, and William—all power players.

So she becomes very worried on the flight to Houston when she starts to feel like she's being pulled out to sea by a rip current. Her head aches in a much more acute way than usual, even with her Lyme. She becomes congested and her breathing is labored. She starts coughing.

How could this be? She cannot get sick now. She's been waiting for two years for an opportunity like this. She tries to distract herself from whatever it is that's pulling her down, but she spends the flight achy and miserable.

When she gets to her hotel room, she lies down on the bed and tries to regroup. But because her Lyme disease has made her so alert to her body's signals, she knows something's up. This isn't just a Lyme flare-up. And if it's a cold, it's a particularly angry one.

Gregg isn't arriving until the following day, and due to some accommodation error, Sophia gets saddled her first night with a roommate, a stranger. It's awkward, though the girl is perfectly nice. But Sophia's so sick, she's wondering how she's going to be able to endure a day of judging, of being on her feet, fully mentally present and able to field intricate questions.

She rallies to attend ISWEEEP's opening ceremonies, an Olympics-like procession. A small group of kids represents each country and Rahul is picked to carry the flag for the U.S. delegation. Seated in the big arena, despite being dogged by illness, Sophia is thrilled to be here. It feels momentous.

Friday, May 5, is public viewing day, when legions of local school kids are brought in to ooh and ahh at the science whizzes. This day always attracts a random assortment of local adults, too, groupies of sorts, scientists and college students. It's a good tune-up for judging.

Gregg arrives, takes one look at Sophia, and they immediately make a run to the local CVS to get her cold medicine and nasal spray.

There's a group outing for all the competitors, but Sophia skips it and heads to her room to try to sleep off the nasty plague ahead of tomorrow's judging.

On Saturday, she dons her science fair attire: a white sheath dress beneath a navy blue blazer and low pumps. She makes her way to her stall in the arena. Upon seeing her, Andy goes out to get her tea.

In addition to being weak, Sophia feels mentally fuzzy, not in the right headspace for what's about to happen. But, mercifully, the judges seem to take great interest in her work, immediately seeing its novelty and potential value. One of them, a veterinarian, is intrigued by the possibility of using Sophia's proposed treatment on dogs with Lyme disease, since it can be very challenging to manage the illness in canines.

"How did you become interested in this project?" ask several of the judges.

"Well, I actually have Lyme disease," Sophia replies. In keeping with her self-made boundary, she offers up the personal nature of her project only when explicitly asked. And naturally the judges are intrigued when they hear her answer.

Other judges take the conversation route. They have a lot of questions but aren't interested in testing Sophia so much as trying to understand all the nuances of her work. Still, she feels she's falling short. It's hard to be confident and persuasive when you're so physically wobbly. Her brain, like Lyme bacteria, feels swathed in a kind of film.

"I think I did average," she says when it's over. "I wasn't sure I was able to get the points across. I couldn't get my thoughts straight."

Because chaperones are barred from the judging sessions, the fair arranges a shopping outing for them. Andy and Gregg decide to hit a local Japanese culture festival, with the goal of sampling as much food as possible.

Their rapport is easy, and during the course of the day, Andy candidly shares with Gregg how he's feeling about his future in Greenwich. He loves what he does, he's had a great run, but he wants to do something bigger and it's looking bleak that it'll happen for him in this town. Also, he's completely fed up with the lack of financial support from the district, having to hold out his tin cup and scrounge from the parents. He can't stand the boom or bust cycles, the unpredictability. It's been wearing him down year by year.

For his part, Gregg thinks this is a travesty and he's willing to say so to the powers that be. Andy's class has been an exceptional experience for Sophia. He's watched his shy, careful daughter develop an unprecedented confidence and take risks. She's developed a dedication to her Lyme disease project that no other aspect of school has inspired in her. To Gregg, if Greenwich lets Andy walk, it'll constitute an egregious loss—one that shouldn't be allowed to happen.

The Sunday awards ceremonies offer no shortage of pageantry. There are dancers and speakers and an elaborately produced live stream of the events.

ISWEEEP gives out honorable mentions and Bronze, Silver, and Gold Awards in the categories of energy, engineering, management and pollution, and health and disease prevention.

Sophia's feeling so sluggish by the time she's seated in the arena, she's not even thinking of winning. She's not merry with anticipation. Instead, she feels like ISWEEEP gave her what she came seeking, which was to compete on a larger scale. She wanted to travel and be given the chance to present her Lyme disease project to a completely new and different audience. And now she just wants to get home and recover.

The first category of winners announced is for energy.

When the Silver Awards come up, the announcer bellows, "Bennett Hawley, United States, Connecticut!" Bennett is one of Andy's sophomores. He designed a wind turbine powered by seawater that, if scaled up, could produce energy for entire municipalities. Bennett had hit a lot of roadblocks with his project but kept at it. Andy's pumped to see him win—and a silver, no less.

It becomes apparent that the results present another parallel with the Olympics—the U.S. is a dominant force, followed by other G20 countries. After that, there are winners from Saudi Arabia, Belarus, Kazakhstan, etc. You've got to cheer these kids on, especially if you imagine the resources they may or may not have. Of particular poignancy are the girls from Arab nations who walk to the podium in their hijabs. Education for girls in some of their home countries is far from automatic or universal, so to see these young women competing in a science fair on the other side of the world makes you think their prospects might be better than imagined.

The remaining five of Andy's kids—Sophia, Dante Grace, William, Sanju, and Rahul—are all in the health and disease prevention category, arguably the most competitive. As the announcer goes through the honorable mentions, none of them are called. Andy and the kids know this means only one thing: either they're on the road to victory or headed home empty-handed.

When the Bronze Awards come up, Dante Grace gets called. She makes her way to the stage to join her fellow winners. Thirty-two Bronze Awards are given.

Next up are the silver winners.

"Sanjeev-Kumar Sathish, United States, Connecticut."

"Sophia Chow, United States, Connecticut."

Sophia hears her name and pauses. She steadies herself and thinks, *Now I have to stand up and make it to the stage.* But as she makes her approach, the corners of her mouth lift into a smile and her eyes take on that feline quality of sweet satisfaction. She manages to get there and finds her place among her fellow silver winners. She looks out from the stage and takes it all in. Whatever it is that's raging inside her body right now can't own this moment.

No, this one belongs to her.

48

Andy

Now that it's May, Andy would normally be entering glide mode. April ended on a particularly high note because Shobhi won her category—and $12,000—at JSHS Nationals out in San Diego, where she and William went to compete. This was yet another first for Andy's kids. Even Shobhi paused to note the magnitude of the accomplishment, pointing out she beat Indrani Das, the girl from New Jersey who won first place and $250,000 at Regeneron STS in March.

Andy's got just one fair left and it's the big one, Intel ISEF, which is being held out in Los Angeles this year. This coincides nicely with him being able to meet his first grandnephew, who lives in Southern California.

Again facing the financial issue of no money to fund Andy's trip, a parent called up Dr. Winters, the school headmaster, and nicely but pointedly complained about there being no money to send Andy to ISEF. Winters said there was, in fact, a fund for such things and made money available to Andy to attend, which Andy's quite pleased about. Andy has never been made aware of any such fund and firmly believes the money magically appeared due to the call.

However, this isn't enough to put him in chillax mode. He's got problems, once again owing to a problematic parent. Andy thought he'd seen the worst of it earlier in the year when a mother whose child failed to advance at a fair sent him an accusatory email that seemed to blame Andy for the outcome. She wrote:

> I doubt that the judges even looked at his work! He needed a headline that proclaimed how it could be applied to selectively kill cancer cells! He wanted to do it all himself so I didn't see the poster until set up. It is something that should have been emphasized. Something that he should have been told.

Andy's no stranger to the sour grapes parent, but he'd never felt so personally indicted before. Historically, aggrieved parents tend to register their complaints in a gentler or even passive-aggressive fashion, seeming not to want to blame Andy explicitly (thereby alienating him) because what good will come of that?

Andy felt that this woman, on the contrary, was standing toe-to-toe with him and spitting in his face. It took him more than a few deep breaths and a free-flowing stream of expletives before he was able to craft a polite response in which he sought to defend himself and also remind her that her child has plenty more opportunities. But she wasn't having it. Without talking to him, she emailed the fair heads.

He'd put that incident behind him, probably thinking that would be the worst of it for this school year. He was wrong. Now, just as he's approaching the academic finish line, he's dealing with the mother of all mothers. So we'll just call her The Mother.

The Mother is actually a fellow GHS teacher whose son is in Andy's class. She's widely known to be a frighteningly intense parent—someone who takes nothing lightly and is particularly obsessed with academic performance and college admissions.

Her son, whom we'll call Z., is a sweet, sunny kid who enjoys theater and music. But The Mother would allow him to pursue these things in college over her dead body. So here at GHS, he's taking some advanced science classes, where, lucky for him, he excels. But as Andy has learned from years of teaching research, kids who are masters at traditional science class don't automatically make for great researchers. The call and response of information-test, information-test has little to do with the ingenuity, creativity, and perseverance required for independent research.

Z. started a project early in the year, but it flamed out. Once that happened, he appeared to lose interest in science research altogether. He used the class to do his homework and horse around with his friends. Somehow, for all her hypervigilance, The Mother lost track of this.

But now, as the year is coming to an end, she's suddenly coming to the realization that Z. has nothing to show for himself. And in a flash, like a shaken bottle of seltzer, she's exploding in a million directions to try to restart his stalled work or put him on a new course altogether. She's gone from an undetectable presence to a constant, lurking one.

The Mother starts by cornering the science research stars and interrogating them about their work. One day after school, she entraps Connor Li and makes him explain in full his project and how he did it. She contacts Olivia via email to set up a meeting, but Olivia ignores her.

The Mother schedules a time to meet with William. She again wants a detailed description of his prior research work and process. She takes notes on everything William says. She kicks things up a notch and offers to hire him to coach and mentor Z. on a project. She specifically wants something that will be "STS-worthy" for next year. To sweeten the deal, she promises William a cut of any prize money—you know, points on the back end.

On May 2, Andy, William, and Sanju are in the lab discussing the incident. Sanju says, "The whole thing is totally creepy." William says he was pretty uncomfortable from the get-go and that it "seemed wrong" right out of the gate. He was taken aback by the grilling.

"She started saying, 'What if we tweak this? What if we tweak that?'" Which made William more distressed. He wasn't in any way offering his own work as a template for Z. Who, by the way, wasn't present for any of these meetings. The Mother has essentially shoved Z. aside to pursue the project on his behalf.

Andy is enraged. He knows he needs to shut down this behavior, but as we've seen multiple times now, such action does not comport with Andy's deep need to preserve his good-guy standing with people—even in cases where they're trying to undermine him, marginalize him, or just plain squash him. He's suspected The Mother of fishy behavior in the past, such as crafting emails as Z., which Andy says he can tell were clearly written by her.

When the research kids back away from The Mother, she turns her high beams on Andy. She starts bringing him piles of scientific papers and asking him a fusillade of questions relating to what's feasible for a project for Z.

But with one fair left in what's been the class's best year ever, Andy has no interest in restarting the engine even if there was time for Z. to accomplish something real, which there isn't. There's less than two months of school left. Besides, as lovable a kid as Z. is, and as much as Andy feels for him living beneath the forcefulness of The Mother, he didn't regroup after his initial project tanked. And this latest maniacal desperation is occurring entirely outside of his desires.

A week after William shares his dealings with The Mother, she walks into the lab after school, scientific papers in hand.

"Let me ask you something," says Andy. "Does Z. even know about this?"

"Yes, but he hasn't read the articles yet," she says.

"So he doesn't know about it," Andy replies.

Andy, trying to appease his colleague in some way, sits down with her to try to get to the bottom of what exactly lights Z.'s fire—even though Z.'s still nowhere to be found.

The Mother has signed him up for a six-week summer science research class at a university to the tune of $8,500. But she develops a fear that the summer program's work isn't sexy enough for the competitive circuit. As some kind of peacekeeping gesture, Andy offers to see if he can get Z. an internship with a researcher in New York City. But he encourages The Mother to be okay with letting Z. focus his efforts on the university summer program. After all, they may as well get their money's worth. He volunteers to read a bunch of papers on the topic Z. will be working on.

But she's got in mind a cancer cell project, all while admitting Z. doesn't like biology.

"Then why are you doing a biology project?" Andy asks her.

Andy tries explaining that the work has to come from Z. and that frankly, he's done for the year after putting in all the hours and staying late and zipping around the country taking kids to fairs.

This somehow only makes The Mother double down on her mission. She starts accosting Andy during every free window he has. She can see his ambivalence and sends him an email saying, "Please don't bail on me now." She tries guilting him by pointing out how much time and energy he's given to other science research kids. When she comes into the lab, she tries to be jovial, but there's a chilling forcefulness about her, an unforgettable look in her eye that suggests she'll do whatever it takes for Z. to produce something noteworthy.

Privately, Andy's like a nuclear power plant on the brink of explosion. He's livid—and also incredulous.

"As great as this year has been with the kids, it's been the absolute worst with the parents," he says. "And she's the worst of all of them."

At one point, she actually asks Andy if he'll open the lab so she can get started doing some of the preliminary work on Z.'s project.

With the constant tension and the unrelenting behavior, you kind of want to print T-shirts that say "Free Z" (and maybe some "Free Mr. B." tees too). Z. is mild in temperament and seems to just step aside, not really having a way to rein in The Mother. Whatever dreams he may or may not have, the sadness is that there doesn't appear to be any space to allow them in. That possibility has been appropriated by her and her dreams.

Since Andy won't confront The Mother, he avoids her. He starts plotting his arrival and departure times to and from school to minimize the chances of running into her. He makes a beeline for his car the minute school lets out. His stomach actually starts bothering him, as the stress of the situation weighs on him.

But her badgering doesn't stop and finally—finally!—he feels he must be completely explicit, since none of his more subtle messages have penetrated. Or if they have, they've been ignored.

Andy crafts a long text message in which he tells her, "I honestly feel like it's time to put this to bed." The New York City internship didn't pan out, so he recommends that they just stick with the university summer program.

He then writes, "It's unfortunate that I have to say this to you, but you've made me extremely uncomfortable with your emails, changes in direction, frequent visits without Z., and last-minute requests, at a time when research activities are essentially done . . . so much so that I don't want to be at school. My class is not about padding a résumé for college, but instead about an authentic learning process. I'm willing to ramp Z. up to what will hopefully be a great experience at [his summer program], once he and the professor have had a chance to chat, but the idea of conducting multiple projects, with or without my guidance, is unreasonable and has little to do with a student following his/her passion. Thanks, Andy."

She never responds.

And with that, he packs his bags and heads to Los Angeles.

As the world's largest science fair, Intel ISEF is quite the spectacle. Held at the Los Angeles Convention Center this year, the five-day affair is part competition, part carnival, part conference, part party. The fair is advertised on every banner and enormous screen around the convention center. Nearly eighteen hundred kids from more than seventy-five countries have descended, and when you walk around the booths, it's clear that the competition is of an entirely different caliber. In order to get here, the kids have had to win a local or regional fair. Many of the projects rival the work that comes out of Andy's class, in terms of sophistication and depth.

He's got two delegations of kids—the ones who are here for having won the CT STEM Fair (Shobhi, Connor, and Agustina) and the ones who earned their berths at CSEF (Luca, Michelle Xiong, Rahul, and Ethan). There are three other Connecticut kids here with CSEF, as well.

However, Ethan isn't here. At the last minute, following some tortured dealings with the CSEF folks, he and his mother, Bonnie, decided the morning he was set to leave that he wouldn't attend. (More on this later.) It was a gut punch felt by everyone. Andy went through the spectrum of emotions, but once he's here in Los Angeles, he's happy to be in the warm weather and sunshine, ready to cheer on his six kids.

In addition to competing, there are cool technology stations where the finalists can tinker during their downtime. Kids congregate around a "virtual reality playground," as well as a station where they can build their own Rube Goldberg machine. Intel ISEF rents out a portion of Universal Studios for a night, so the kids get the amusement park all to themselves. And there are lots of panels and sessions for kids and grown-ups alike.

During the opening ceremonies, there's a grand parade of all the participating nations. Many of the kids from abroad wear traditional or local attire. There are Korean girls in hanbok dresses and Saudi boys in thobes and keffiyehs.

The kids are wide-eyed and awed by it all. They know that by mere virtue of being here, they've ascended to the big leagues.

One of the events that has the biggest buzz is a panel featuring particularly enterprising and unusually high-achieving Intel ISEF alumni. These are entrepreneurs who started and sold companies in their twenties and in many cases have taken a less traditional path.

These alums aren't exactly household names, but some are very well known among the science fair crowd. Geoffrey Woo, twenty-eight, is the CEO of HVMN, a company that makes and sells noo-tropics, purported cognition-enhancing compounds. His company's mission statement says, "We believe that the human is a system that can be quantified, optimized, and upgraded."

One of Andy's kids, Luca Barcelo, gushes that he was able to get a photo with Woo. "Oh, my god. I couldn't believe it," he says.

Yet another panelist is Taylor Wilson, who became famous after he performed nuclear fusion at the age of fourteen on a fusor he'd built in his parents' garage. In 2012, he received a Thiel Fellowship, a $100,000, two-year scholarship given to kids in exchange for forgo-ing college for those two years.

Sheel Tyle, a twenty-six-year-old venture capitalist who graduated from Stanford at age nineteen, tells the rapt audience, "ISEF made me realize you don't have to be too old to create change. When people say you're the innovators of tomorrow, they're wrong. You're the in-novators of today."

Tyle tells the group to embrace failure. "I've had companies that have failed. Failure should be celebrated. If you don't fail, it means you're not trying to do something big enough."

It's a shame that Ethan's not here, because these truly are his peo-ple. It would be impossible for him not to feel like an anomaly in Greenwich, or most places, but these panelists are proof that an ac-tual Ethan peer group exists.

Judging day is sealed off to the public. More than one thousand volunteer judges are on-site. Prem Subramaniam, Rahul's dad, had applied to be a judge but recused himself once Rahul became a final-ist. There are twenty-two categories that span every corner of sci-ence, math, and engineering. Each finalist is visited by about eight judges.

Connor Li is the only senior from Andy's class. He's already ac-cepted a spot at Cornell, to study engineering. So he can truly just savor the moment here at ISEF without thinking about what it will mean.

With the performance pressure eased, Connor gets into some

meaty conversations with the judges, who he says asked him more pointed questions than he'd experienced at other fairs.

"ISEF definitely had more specialized judges that had backgrounds in the category of my project," he says, adding that he appreciated the engaging and higher-level discussions.

During the public viewing day, Shobhi is highly amused when she gets bombarded by South Asian moms who seek her out to ask what they can do to get their kids to be more like her. As if that's answerable.

Her category, computational biology, seems to be one of the toughest this year. There are many projects similar to hers—that is, kids who have built algorithms around cancer treatment or detection.

Rahul and his Zika project are competing in the category of microbiology. For other competitions this year, Rahul has listed his father, Prem, as his mentor. But for Intel ISEF, he has listed Andy. Rahul's project garners a lot of attention from judges and people who come to the public viewing. Like Olivia's Ebola diagnostic, Rahul's project has a purported intervention for a frightening, headline-generating disease.

After a week that is both exhilarating and exhausting (Shobhi actually falls asleep on a ride while at Universal Studios), for the most part, Andy's kids don't seem to be focused on winning anything. It's as if they've gotten their fill just being here. Their casual approach might also stem from the fact that they've been humbled by the significant jump in competition level. It's unquestionably easier for them to stand out in Connecticut. Here, they're among equals.

On Thursday evening, everyone files into the massive main arena and takes their seats for the awards. For these prizes, there are monetary awards given to those who place between first and fourth in the twenty-two categories, ranging from $500 for fourth to $3,000 for first. (There are multiple winners for each place, so there could be, say, seven kids who win fourth place in animal sciences, five who win third, etc.) Then there's one kid who gets named best in category who takes home an additional $5,000, and their school and local fair each get $1,000.

And then, from the pool of best in category winners, the judges select the top three to receive $75,000, $50,000 and $50,000, respectively.

Sitting in the auditorium, Andy's got his Nikon ready. "The goal is to get to the stage, no matter what it is," he says. He's had several kids make it up there over the years, including Olivia, William, and Ethan.

The winners across all twenty-two categories are called by their places, starting with fourth-place winners and then moving on up. As their names are announced (most of it is prerecorded in, oddly, a British accent), video cameras capture the kids walking up the aisles and heading to the stage to thumping techno music. Their images are projected on towering screens. Some of the kids twirl and do little jigs, raise their arms up, fist-pump. It's a brisk, clean operation that moves with alacrity.

The fourth-place winners are up first. A few minutes in, the pre-recorded voice says, "From Riverside, Connecticut, Shobhita Sundaram!" Andy whoops and cheers. This is actually Shobhi's second trip to the big stage, as she won a specialty award the night before.

As the image of Shobhi walking up the aisle flashes on the screens that are as tall as a building story, she looks both pleased and stunned.

Later, though, there does seem to be some disappointment that she didn't place higher.

"Fourth place is good, right?" she asks, earnestly and seekingly, in a rare glimpse of insecurity. Andy's a little surprised, too, that she wasn't closer to the top, though he'd never betray that.

During the third-place announcements, none of Andy's kids are called.

When the second-place winners' names start booming out over the arena, Andy's ears prick up when he hears, "From Greenwich, Connecticut, Luca Barcelo!" Luca developed a cool, cheap paper-based device that tests for nitrates in bodies of water. You fill a special cup, take a photo with your phone, and based on the color change, you can find out whether nitrates are present and if so, in what quantities. The data are uploaded to a crowd-sourced website so people can see where the greatest levels of nitrates are present. He's had exploratory conversations with the EPA about his project.

Andy's pumped. Luca's a hard worker who's unafraid to teach himself things his project requires, like building a website.

Sitting in the audience, Andy ticks through his remaining students: Connor, Agustina, Rahul, and Michelle Xiong. Hmmm, all great projects. But are they first prize–worthy? Impossible to say.

For the first place and top awards, the prerecorded British lady is replaced by a live human, a local newswoman.

"In the category of microbiology, from Cos Cob, Connecticut, Rahul Subramaniam!" she bellows, really punching the end of his last name.

As soon as Andy hears "Cos Cob, Connecticut," he starts giggling, clapping, and yelling, "Woohoo!"

But it's not over yet.

Rahul goes on to win best in category for microbiology, bringing his winnings to $8,000. He also wins an all-expense-paid research trip to India.

Andy's never had a kid win the top prize in his category before. None of his kids have ever made it this far.

When the top three prizes are announced, second place and $50,000 go to Amber Yang, the Regeneron finalist who developed a way to track space debris. When the top three winners take the stage, confetti tumbles down and the applause goes supersonic.

The GHS kids eventually make their way to the floor and meet up, huge amounts of congratulations all around. Rahul, usually a half smiler, has got a full-face grin on display. The kids rib him a little for supposedly jumping up to take the best in category award before his name was actually announced. But it's good spirited and there's an air of deep satisfaction among the group.

Prem sends Andy a congratulatory text. Andy takes the high road and responds in kind. He's still unable to forget or completely forgive Prem's transgression this year, but he's thrilled for Rahul.

And tonight, the focus is on the kids. They walk out of the convention center into a warm, breezy, perfect L.A. night in search of some celebratory ice cream.

Andy doesn't often pause to tally the wins, but when he thinks about it, this year has superseded them all—in terms of kids advancing, winning, earning prize money. Between Regeneron STS, CSEF, CT STEM, JSHS, ISWEEEP, and now Intel ISEF, his kids have absolutely dominated. Greenwich High has been all over the science fair circuit map.

But this year will also endure in his memory because of the kids themselves. What he'll mostly remember are the moments that happened far from the awards stage—the ten P.M. night when Collin's project came together, working with Romano in the woodshop to saw those windows into his test tubes, that cozy winter day with his hardworking female superstars. He'll remember some of the tears over various teenage crises, a clutch last-minute prom date he helped broker, his sense that he may never again have a class like this year's seniors, with whom he formed an especially tight bond.

On the plane back home, Andy settles into his seat and thinks through it all. Yes, he wants to move up, expand his role, take on something bigger. But no matter what form a new job may or may not take, he can't leave the kids. They are the ticket to his happiness. They've become such a central part of his identity, and most critically, they've given him that elusive thing so many of us seek: a tangible sense of mattering in this world.

49

Ethan

Ethan quickly adjusted to life in San Antonio. Bonnie, a master of logistics and planning, tended to all the details. Ethan briefly wanted a pickup truck, which are abundant in Texas. But Bonnie and Keith purchased him a secondhand BMW sedan instead. Bonnie rented him a two-bedroom apartment in a nice complex so that she and Keith can stay there for extended periods. In terms of her presence, it's a fine line to balance. She wants to be there and provide Ethan support, but since his days now consist of turning up at the Southwest Research Institute in the morning and leaving in the evening, there's not much for her to do. So she spends stretches in San Antonio and then travels back to Connecticut to see Keith.

For all of his star status, or perhaps because of it, when Ethan wins one of CSEF's Intel ISEF berths at the fair this year, a strange and not completely explicable dynamic emerges between the Noveks and the fair folks.

CSEF insists that all of the Intel ISEF kids travel together as a delegation—partially for liability reasons. But since Ethan's living in Texas, he wants to fly to L.A. directly from there. This sticks in the craw of the CSEF heads, despite the fact that it makes perfect sense. The Noveks try to sand things over by saying they'll purchase Ethan's plane ticket and spare the CSEF people any additional hassle of having to make separate travel arrangements for him.

Bonnie and Ethan likely don't help matters by saying that Ethan needs his own room (as a senior, he is entitled to one anyway), with the explanation that Ethan needs to make business calls during his downtime and he doesn't want to bother a roommate or be bothered himself. Ethan's schedule is, in fact, jammed with calls, as he's constantly networking for possible business or partnership opportunities, but certainly he can wander off to a corner of a hotel and make do.

The CSEF chaperone finds unacceptable the idea that Ethan would be managing his business affairs while at ISEF and sends him chastising emails warning him not to do so.

So in the two months between the time Ethan wins his Intel ISEF berth and the morning of his departure for the fair, an antagonistic relationship sets in, a petty power struggle of sorts.

Ethan says of the main chaperone, "The woman who runs the trip, she was a former social worker and she was treating me as someone who needed discipline."

The morning he was slated to leave for L.A., CSEF director Sandy Müller called him to confirm whether he was going, though he'd not given any indication that his attendance was even in question. This felt taunting to Ethan and Bonnie.

Following Sandy's call, Bonnie plainly stated to Ethan, "I don't want you to go." She had a bad feeling about it. For his part, Ethan was torn. He knew not turning up was a big deal and genuinely didn't want to "wreck relations" over it. Setting aside the strain between him and the fair heads, he had tons of work in regard to the construction of his prototype. He called his older brother, Chad, for advice. Chad stripped the whole sordid affair down to the fact that if Ethan might not be happy and if he felt the CSEF folks were tying him in knots, he should just skip it. So he did.

Ethan called Andy and gave him the news. Andy took an incredibly measured tack, as he's very mindful of not trying to twist kids' arms or apply undue pressure, since most of them operate with extraordinary amounts of stress from other adults in their lives. Andy also likely treated Ethan more gingerly because their dynamic was different. Ethan hadn't been part of the class since junior year, and since he was leading much more of an adult life, it felt more natural to speak to him as a peer. Of course, the thought that Ethan would scrap a valuable spot that could have gone to another eager and deserving kid weighed on Andy, but he set that aside.

"I told him, 'Look, you've jumped ahead in life in a lot of ways. But I can tell you, the one thing about work—it's always going to be there, there's always going to be more. So you might not want to miss out on this for work.'"

"They were threatening to make my life miserable," says Ethan, ratcheting the drama up to full teenager level. "I hit this point, I said, screw this, I'm not going to Intel." But in scrapping his trip to Intel, he also lost an opportunity that's increasingly rare for him, which is social contact with people his own age.

"I definitely missed out on something great," he says of Intel. "I have a lot of friends there; I would have met a lot of people."

However, he says that the week he should have been at Intel ISEF was a really productive one for him in San Antonio.

At the time of this writing, Ethan is living in Norway and working with a power plant to test his ever-evolving technology on a larger scale. At SwRI, he completed construction of a small pilot demonstration unit that successfully uses his carbon capture technology. Among its many benefits is the cost. Carbon capture is often measured in dollars per ton—that is, the cost of capturing a single ton of carbon dioxide. The current industry standard is $70 to $100 per ton. Ethan's technology comes in at $8 a ton. For the Carbon XPRIZE, the necessary threshold to advance in the competition is 60 kilograms of carbon dioxide a day. Ethan says his SwRI unit far surpasses that threshold.

"What I have now is perfectly fine for proof of concept," he says.

The Noveks have employed a team of intellectual property attorneys to keep up with his various developments. The lawyers advised Ethan to withdraw from Carbon XPRIZE because the competition does not align with his long-term business objectives. Scientifically, the competition's emphasis departs from Ethan's, so he decided to follow his lawyers' advice.

Ethan's still constantly connecting and networking with prospective collaborators and investors. High school, which isn't yet over, seems like a distant memory.

In May, following the Intel ISEF debacle, Ethan's slated to be one of the featured speakers at the Connecticut Academy of Science and Engineering (CASE) Awards. It's a prestigious evening and the Noveks are genuinely happy Ethan's been asked to speak.

Andy attends, despite the fact that by this time of year, he's maxed out on awards and galas and any evening event involving an overcooked chicken entrée alongside limp vegetables.

The CSEF folks are present as well. Everyone smiles and hugs and no one betrays any grudges. Though shortly after, Sandy emails Bonnie seeking more than six hundred dollars for losses incurred over Ethan's no-show. Bonnie and Ethan are apoplectic and insulted because they feel the CSEF people essentially forced Ethan out. CSEF feels that given its tightrope budget, they can't afford to eat the costs of a no-show. Bonnie forwards the email to Andy, seeming to want some guidance. Andy wants no part of the conflict.

In the end, Bonnie writes the check.

At the end of the school year, though he's living full-time in San Antonio, Ethan's facing one last niggling matter from back home—

graduating from high school. Greenwich High determines that he failed to complete the wellness requirements to graduate. Ethan held up his end of the absentee student concession given to him by the school in all respects except this one. He probably would have had to work out something to check the wellness box—showing he was taking a fitness class in San Antonio or some such. But it never happened. And insofar as he thought about it, he may have assumed the school wouldn't hold him to it.

But it does. It seems that from the school's perspective, Ethan got a special deal to pursue his ventures as he wished, on the condition he complete all the necessary requirements. He didn't do it. So, no diploma.

Bonnie and Keith meet with administrators but discussions don't produce a satisfactory resolution. They grow concerned that Yale might withdraw Ethan's spot if he doesn't have a high school diploma.

On June 20, Ethan does not walk with his class or attend graduation. Bonnie feels burned by the whole thing and that her son has more than shown his graduation-worthy academic achievement. The wellness credits seem so petty, so shortsighted. Despite having missed the last year and a half of school, the football games, the proms, etc., it hurts that he's not among his classmates for this one ritual. It upsets Bonnie far more than Ethan. "It's just a very difficult day," she writes in a text message. And who can blame her? Even though high school was never Ethan's natural habitat, most any parent wants to see her child participate in a basic rite of passage with his or her peers.

And then a weird thing happens. A few days after graduation, the school mails Ethan his diploma.

"They ended up just kind of waiving it," says Ethan. "They backed down, sent the diploma to me, and I was on the dean's list. I'm so happy it worked out fine, but what does it really matter to me in the scheme of things?"

He feels it taught him a different lesson—certainly one that departs from the message the school was perhaps hoping to send.

"They always tell you, be different, be yourself. When you actually live by those motivational things, they don't reward it.

"In life, there's the standard experience and the lives you could have. High school is way overrated. If you look at people who work hard in high school, they have to spend a lot of time on things they don't like doing. I get to spend my time doing the things I want to be doing."

50

Danny

When junior prom rolled around, Danny didn't make it. It was likely a combination of apathy, lethargy, and not wanting to have to prospect around for a date.

But this year, he's committed to going, and it's a good thing because the prom has become a source of much discussion at the castle. This time, it seems whiffing on the prom is unacceptable—and everyone is urging him to get it together.

Which he does. After some back-channeling by others, he asks a junior who gladly accepts. And then Danny and his date participate in all the prom hoopla—the pre-prom get-together for photos, the prom, the after-parties. Danny says he had a great time. Everyone seems content with the outcome.

But then Danny gets himself into a predicament at which even he can only marvel for its lunacy. "It's so *Curb*," he says, referring to one of his all-time-favorite shows, *Curb Your Enthusiasm*, where Larry David ends up in any number of excruciatingly awkward scenarios.

Leah's away on business and Danny's driving around town after school one day. He's on a street that runs parallel to Greenwich Ave.

He's stopped at a light and starts to move forward when it turns green but doesn't look at all sides of oncoming traffic and gets into a fender bender. He's fine, and so is the other driver.

But what he cannot believe is who the other driver happens to be. Danny hits none other than the elderly father of Greenwich's police chief. He's mortified, instantly fearful that hitting the father of the town police chief will carry with it some extreme punishment. If stockades were still in use, he thinks he'd be sent to them. And remember, his little crack-up comes on the heels of endless joking at the castle about what a bad driver he is. Danny shut out all the banter about his driving skills because in his own mind he's very safe and competent behind the wheel. Which could be the case. It might just be that he has very bad luck.

At GHS, the seniors actually get a furlough for the month of June to pursue independent projects. Many of the projects are flimsy, at best. This year, one senior's project consists of going around town and reviewing donuts. Some kids intern at local businesses. William and his friends are building a videogame. At the very end of the year, there's a fair where all the projects are displayed, but there are no grades and the whole thing seems like a means of shooing the pesky seniors off the premises for a month.

Danny and his friends are building a go-kart with an internal combustion engine. Henry Dowling and his posse are also building a go-kart, but they're planning to buy all their parts, whereas Danny and his cohort are intent on making all theirs.

They encounter a series of problems from the start. They were going to have some parts custom-made in a mill, but the mill can't do the job in the needed time frame. There are design issues and disagreements on best methods. And much of the work requires outsourcing that isn't coming together.

In the end, their go-kart is 3-D printed (it takes thirty-nine hours to print), but Danny calculates that due to several design issues, the temperature from the engine will actually melt the car's exterior in twenty-seven minutes—making it a single-use go-kart.

While Danny is working on his go-kart, he also spends the last week of school turning up at the lab every single day, true to his promise to finish his mushroom battery. The week amounts to his most consistent work stretch on the battery all year. When the kids and Mr. B. express their shock and dismay at Danny's presence, he says, "I said I was going to finish it and I'm a man of my word."

And guess what? His battery sort of works. He charges it up and it reaches a voltage of 2.6, a completely respectable level of energy output. One could argue that it's somewhat miraculous that it works at all.

The problem is that after ten cycles—that is, recharges—the voltage drops to 0.8. Batteries are meant to last hundreds of recharging cycles.

"The fact that it's gone down so rapidly, this has lasted fifteen cycles at most, I think it's because the lithium I used was pre-used. I think that's one aspect to the fact that it's degraded so fast," he says.

Had he assembled and tested the battery back in the fall, he could have dug in and figured out its flaws, tinkered with the configuration, and maybe gotten it to endure more cycles. But it's too late for that now.

At graduation, someone snaps an epic photo of Danny yawning during the ceremony, which he proceeds to make his profile photo on Facebook.

The photo didn't actually capture his mood about the milestone moment.

"I was nervous the entire time," he says. "The future's coming pretty fast, I'm afraid I might have missed something along the way."

Three days after graduation, he's come to the city to Rare, his favorite burger joint—the same place where, nine months ago, he and Andy gorged themselves into food comas following their day at Mount Sinai.

In characterizing his year in science research, he doesn't hesitate to offer his own self-assessment as to why he couldn't quite motivate.

"Senioritis," he says. "One hundred percent being a senior. I look around at the other seniors there and they all had a reason why they had to keep going. Like Yin, for example."

He then goes on a winding explanation of his thought process. For one thing, he says, the early admittance to Harvard didn't help in terms of lighting any fires.

"If I had not gotten in early, who knows? I might have gone balls to the wall with it and tried to finish it all," he says. "I was a lazy senior, I was the stereotypical lazy senior.

"In the end, I did get it done, I just didn't care enough about the fair season. For me it's never been about the fairs." Which seems legit.

He says that when he interned at Rockefeller University the previous summer, there was a similar vibe to the class, in that no one was haranguing the interns to work; they had to be self-driving vehicles. There, he says, he stayed until seven or eight every night—without the end zone of a science fair. His determination came from within.

As for his mushroom project, he says, "I never loved it." His junior year project, the one that would have required him to work with human blood samples, which every institution denied him, stoked his internal fire.

So, about that illicit pizza delivery every "E" day, the feat for which Danny earned considerable cachet . . . how *was* he able to sneak in a steaming stack of pizza boxes past the security guards? What was his trick?

Now that he's graduated, Harvard bound, and, god help us, about to embark on a summer road trip to Canada, he's willing to give up his trade secrets.

The pizza delivery started back when Ethan was in the class. The two had been great friends since middle school. At first, Ethan would locate a door where no security guard was present and text Danny where to direct his nanny to drop the pizza.

"The security guards inevitably found out about this. Of course, I wouldn't care. I would just keep on doing what I was doing until eventually I'd had enough with that route. It would get annoying because I'd have to go through a different route every day.

"What I ended up doing was I would literally just carry it right by them. And they knew it and they would never stop me. They'd stop everyone else bringing food in, but they'd never stop me," he says.

"I figured that they had to have some reason for it. I don't know what it was. I walked right by and upstairs." All the other kids trying to sneak pancakes and burritos from the cafeteria into the science wing would get busted and turned away. But somehow Danny's playing it cool cast some inexplicable spell over the otherwise vigilant security staff.

So Pizza Man's method wasn't so much an ingenious workaround; there was no secret hatch or drop box. It was merely that Danny acted as if the pizza was his right.

And it worked.

In the end, what did Danny get out of the class? To him, the value wasn't in toiling away on his own working on a project on which he never had more than a fleeting crush.

The value was the domain it gave him, the social scaffolding, the persona, and in many ways, thanks to becoming Pizza Man, the role of hero.

"At the end of the day, I like the science aspect of it. This past year, I got to talk to people about the science. I worked in a virology lab, I talked a lot with Rahul about his Zika project because I worked a little bit with Zika when I was at the lab.

"I mean, the people in it are the best people, some of the best people I've met. My friends I knew going into it, I became much closer with. I mean, I wouldn't trade that experience for the world."

51

William

While at ISWEEEP back in early May, William and Sanju were sitting in their hotel room during some downtime. William was on his computer while Sanju was talking to him.

William got a Facebook message from a female friend, someone he met at ISWEEEP the previous year, a girl named Mary from New Hampshire. She sent him enthusiastic congrats on being named a Presidential Scholar—one of the nation's most prestigious accolades for high school students. Awarded by the Department of Education, scholars are named from all fifty states and invited to Washington for a three-day stay in the nation's capital, to sightsee, meet with elected officials, and attend an awards ceremony.

When William got the kudos from his friend, the magnitude of the news didn't quite register with him.

"I had totally forgotten about it," says William, who had applied to an array of competitions and scholarships. He went to the website, scrolled through, and found that he was one of three scholars from Connecticut.

Whether it was because he's got some scar tissue from some of the year's considerable letdowns or because he thought the Presidential Scholar award was a long shot, he was in complete disbelief. He sat in his hotel room, unable to speak. Seeing the look of bewilderment on his face, Sanju said, "What's wrong with you?"

He said to Sanju, "I just found out I'm a Presidential Scholar." Sanju offered his quick congrats and just kept talking.

William was able to name a distinguished teacher on his application. He chose Andy.

Once William was back from ISWEEEP, Andy congratulated him and said, "Thanks, buddy. I appreciate it. I wasn't able to get there on my own," pointing out his failed bid at a Presidential Award. Andy seemed genuinely touched, not because of the public recognition, but because it was William's definitive gesture that symbolized all they've lived through over the past three years. Science research has been such a pillar of William's high school career, of his identity—and to share the glory with Andy was to acknowledge all of this.

The Yins invite Andy to the Presidential Scholar ceremony. It's June 18, Father's Day. Andy would like to be with his family, but he can tell how seriously happy and proud William is. He decides to attend because after all, this is a once in a lifetime event for William—and in light of this year's lumps, this is a very satisfying and triumphant way to end both the year and their time together.

The Yins drive down to D.C. Andy takes Amtrak and arrives an hour or so before the ceremony, which is held at the Andrew Mellon Auditorium, a grand Classical Revival building in the Federal Triangle.

Andy takes a seat midway back in the audience. As seven thirty draws near, family members and friends start filing in. Once everyone is seated, the 161 Presidential Scholars process into the auditorium.

The ceremony itself is noticeably lacking in any frills or even basic touches. There's no program with the names and bios of the scholars, merely an 8½″ × 11″ piece of paper perfunctorily listing their names and states. There's no video montage of the kids giving people a sense of the scope of their accomplishments. One of the more intelligent things the major science fairs do is document the kids via video while they're at the fairs. They then showcase the videos at the closing ceremonies, giving everyone a window into who these kids actually are.

The Presidential Scholars ceremony does, in fact, show one video. It's from the controversial secretary of education, Betsy DeVos, who doesn't make a live appearance at the awards. Instead she delivers a cringeworthy greeting in which she wears an apparently immovable frozen half smile and her eyes can be seen darting horizontally as she reads from a teleprompter throughout. Everything about it—the lighting, the too-tight facial close-up, her forced way of speaking—is like a send-up of a horror film.

Following the video, a couple of people deliver predictable remarks about youth, achievement, the future—all the usual items on the menu.

The one bright spot in the totally lackluster ceremony is the keynote speech by Congressman Jamie Raskin, a Maryland Democrat. Raskin's a funny, self-effacing, energetic speaker, and he elevates the gathering, giving it the greater sense of purpose it deserves.

Raskin himself was named a Presidential Scholar as a high school senior. It was here that he met the guy who would become his lifelong best friend, the best man at his wedding, his compadre. After meeting his friend at the three-day affair in Washington, D.C., he traveled across the country to visit him, a formative experience for a teenager.

Raskin asks the scholars, "Has anyone been in love?" A few kids timidly raise their hands. He then implores them to fall in love—with this country. He tells them to connect with fellow scholars from other states, to visit one another, and in so doing, to open their eyes and hearts and take in what America has to offer, to sample its many landscapes and talk to as many people as they can, to seek out the best of what our nation has to offer.

Such an entreaty could be treacly, but during these most divisive

of times, it's an urgent, genuine call for kids on the cusp of independence to resist any urges to allow division and cynicism to become
their default view of America. It's an affecting speech.

When the moment comes for the kids to receive their medallions,
they are called up alphabetically, so William is at the end of the pack.
He walks confidently across the stage, lowers his head to receive his
medal, and then poses for a photo with one of the award heads. It's
over in seconds.

The ceremony wraps and in keeping with the no-frills aspect to the
affair, despite the fact that it started at seven thirty P.M. and it's now
after nine, there are no refreshments, not a glass of water or an hors
d'oeuvre to be seen. Families have traveled from all over the United
States for this prestigious award, but they don't even get a crudité or
crostini. (They are, however, treated the following night to a performance at the Kennedy Center by Presidential Scholars in the Arts.)

The families and kids linger in the auditorium, posing for photos.
A Chinese film crew is documenting William for some program in
China, and they're aggressively tailing him and shoving a mic under
his chin. William's nonchalant about the whole thing, appearing to
have mastered the art of ignoring the media. Some of the other kids
stand back, ogling and giggling at the scene.

The Yins and Andy gather around William. The Chinese reporter
taps Andy and starts asking him questions about the science research
program and about William.

"They become good scientists, but good people, as well," Andy
tells her. "With William, I think 'scientist' is too . . . narrow. He's more
of an innovator, a creator."

William says, "I really have to thank Mr. B. for all he's done, the
countless evenings, holidays, breaks, snow days."

Andy chimes in, "Nights at Sacred Heart," where they once had to
work because they needed certain equipment. "I could think of nothing better to do with my time," he says, grinning.

William and Andy give each other a hug. Andy's face gets serious,
a bit sentimental.

"Congrats," says Andy. "I'm proud of you. I really am. You deserved this."

During his parting comments to the film crew, Andy says, "Someday, I'm going to say I knew William Yin. I helped him a little on this big journey he was on."

With all the pageantry and drama of college acceptances behind them, there is one bit of business left among the Greenwich High seniors: Who will be the valedictorian of the Class of 2017?

Many assume Yin will be number one. They say it's just not humanly possible for someone to have bested him, based on his coursework alone. Others are not so certain, noting that the school's complex grading and weighting system, in combination with such an exemplary senior class, means nothing is certain. There could be some dark horse who's going to come out ahead.

Honors and AP classes receive more weight, but there are some honors elective classes that are considered easy and could theoretically boost your standing. Kids are eager to point out all the ways the system can be gamed and that the less deserving could go about their course selection in a crafty manner and surge ahead.

Greenwich High does not assign classwide rankings. For the purposes of valedictorian and salutatorian, they tabulate the top two students in the class and that's it. And the calculation is done just weeks before graduation, to try to minimize uproar. There's already enough gladiator-like competition in town, so the administration doesn't see any benefit to giving the kids one more thing over which they can try to clobber one another.

A perfect GPA in the GHS system is a 5.3, but it's technically impossible to achieve because of AP course restrictions placed on classes kids take during freshman year.

The prior year, however, was laughable. In 2016, the school named eight salutatorians, claiming there was an eight-way second-place tie—enough salutatorians to hold a water polo match. Since there's no time or interest in eight salutatorian speeches, each number two got up and said a few sentences. Many people thought this was ludicrous, the ultimate display of "every kid is a special flower" culture. In the words of chemistry teacher Shirley Barban, "Seriously? Take it out one more decimal place and pick someone."

So this year, that's what the school is doing. The GPAs will be rounded to the hundredth place instead of the tenth place.

A few weeks ahead of the announcement, a crew of curious seniors heads to the guidance office to try to sleuth out the answer. They somehow convince one of the secretaries to look up the info in a database and confirm or deny whether Yin is first.

No, she tells them, William Yin is not. A female is in first place.

When the curious crew reports its findings, they seem to know it's bad form to be gleeful. It's hard to tell if their smirks are due to the fact that they've unearthed this little gem of a scoop or because of the news itself. Maybe it's both.

On May 30, William's eighteenth birthday, he and Verna go to Starbucks, where he gets a text from his mother telling him to call Dr. Winters. He does as told.

The headmaster says, "Hi, William. I have some news. You're probably not surprised. It's what you were hoping for. Oh, and you should probably tell your mom." Winters had called Wendy and asked her to ping William.

William had been gently poking around trying to find out whether he was going to get the valedictory spot. As recently as a couple of weeks ago, he asked Winters, who declined to give him an answer.

The news is undoubtedly one of William's best birthday gifts ever. He's elated. Besides Harvard, this is the one accomplishment he so very badly wanted.

Given his long trail of achievements, it's intriguing that William attaches such importance to being class valedictorian. Earlier in the year, he sheepishly admitted that it was just a "cool thing," part of the school's history, a prestigious title. And he also really enjoys public speaking. Being valedictorian would give him a considerable platform. Plus, unlike his science fair presentations and other speeches, this one won't be a competition. It's a public moment of pure triumph, a culmination of his four years of domination.

Once he gets the news, he doesn't try to play it cool. He immediately tells Andy. And without hesitation, he says, "I'm really happy."

As for the misinformation, the curious crew is stumped. They don't know whether the secretary purposely threw them off the scent or if the final exam grades changed the result. Theories abound, but whatever happened, there is no schadenfreude fest to be had.

Graduation is Tuesday, June 20, at six P.M., on the football field. Should it rain, the ceremonies will be held inside at four P.M.

William doesn't seem to be able to un-William himself, even with this. He initially plans to fly back from the Presidential Scholar awards in Washington that day, on a flight that lands in Westchester (about thirty minutes from Greenwich) around two thirty P.M., leaving a gaping opportunity for him to miss graduation. But on Monday, there's a full schedule of activities that William doesn't want to miss, including a White House visit. Plus, Tuesday is the day earmarked for scholars to meet with their elected officials and William wants in on that, too.

Everyone in his orbit over the age of eighteen thinks his proposed itinerary is a woefully bad idea. His parents, Andy, anyone who hears his cockamamie plan. After the Presidential Scholar ceremony, Mark Yin looks positively weak, like he's not even fully breathing, at the mention of the Westchester flight arriving at two thirty.

"William, you can't miss graduation," he says. "You cannot risk missing it."

After three years of torment from William's alternate theories about time-space and what is possible, it gives Andy great pleasure to offer Mark a better, more workable plan.

"Can I make a suggestion?" says Andy.

"Please," says Mark. "What do you think William should do?"

"Since William doesn't want to miss out on the Monday stuff, why don't you let him finish out the day and then you all drive home late Monday night. Just toss him in the minivan and drive back to Connecticut. Forget the flight and Westchester," Andy says, tying up his idea with a bow. "He can sleep in the car and you will definitely be back in time for graduation. It doesn't matter how late you get in, he'll be there the next day with plenty of time."

"Ah, you know, that's a great idea," says Mark. "That's so much better. That's what we'll do."

William checks back into the conversation and Andy takes the fall for Team Adults with Life Experience and Rational Thought.

"William, forget the Capitol Hill visits on Tuesday," Andy says. "I mean, come on, what are they really going to say to you? Anything important? I don't think so. Stay tomorrow, go to the White House, do your thing, and then drive home with your parents on Monday night. This way, there's no chance of you missing graduation. Because if you miss that . . ."

William has to process this new plan but comes around quickly, even conceding that it will save him stress and anxiety. The color has come back into Mark's face.

On Monday, William goes with the group to the White House for the privilege of meeting Melania Trump, a kind of science experiment in her own right. She takes a formal photo with the scholars, which she tweets out.

The Yins hit the road at one A.M. William kept them waiting for an hour as he was saying his goodbyes. He has not yet written his valedictory speech. Near the end of the car ride, he writes the intro paragraph. They pull into their driveway at ten A.M. and William then sits down and churns out the speech in one go.

Technically, William hasn't completed his line-item graduation requirements. He hasn't turned in all his books, his mellophone and French horn, or his Chromebook. He doesn't imagine he'll be barred from the ceremonies over these things, but it's on his mind.

He says, "It's sort of funny, but it's a habit I've got to work on. This seems to happen quite a bit."

June 20, the last day of spring, is a banner day for an outdoor graduation. The temps are in the high seventies, comfortable for enjoying an afternoon in the bleachers. It's daunting to think of a graduation of 650 kids. But Greenwich has fine-tuned the ceremony to make sure it runs no longer than two hours. The speeches are few and brief and while they do call the name of every graduate, they have four circles of kids moving at a time, so they push through with speed and efficiency.

Teachers are not required to attend graduation, and it's somewhat pathetic that of a faculty of roughly 250, about 20 march in the ceremony. Those who do gather for a catered dinner in the teachers' dining room beforehand and talk about how shameful it is that their colleagues can't be bothered to show up. They all point out how the ceremony moves so quickly, it's not as if they're giving up an entire evening. And even if they were, so what? It's graduation, it's their kids' major milestone, and it's one day out of the year. All that lies ahead is their summer off.

Shortly after five P.M., the high school band strikes up what has to be the fastest, most up-tempo rendition of "Pomp and Circumstance" ever played. It's completely ridiculous, sounding like a record on the wrong speed. And because it's so brisk, the administrators, faculty, and kids look as if they're race-walking onto the football field. It would be hard enough to find your graduate in the sea of red and white caps and gowns, but now it's impossible because they're in a fast-moving blur, as if they might not get their diplomas if they walk at a normal pace. While the school is certainly making good on its reputation as the paragon of the down and dirty graduation, this entrance looks like a parody, a comedic sketch about a ceremony in a town where everyone's time is painfully precious.

In just a few minutes, 650 kids have taken their seats and the commencement has commenced. Dr. Winters and then the interim superintendent stand at the lectern offering the classic graduation tropes. The salutatorian gives her speech. And then it's William's turn.

He steps up to the microphone and opens with the confession that when he entered high school, he wished and believed he'd write the great American novel. He envisioned his tome rivaling those of Steinbeck, Morrison, and Hemingway. He admits he never wrote the book; he never made it past the first chapter. But what he gleaned from the experience and what becomes the theme of his valedictorian speech is fearlessness.

William says,

> For all of us, there have been moments when it felt as if all was lost, as if we couldn't imagine the sun rising the very next day. For me, one moment in particular stands out in my mind. For two weeks of my junior year, my life became engulfed by a single, seemingly trivial problem with my research, involving passing drugs through a pigskin membrane. Every day was another day in which I felt as if I were drifting further and further away from what I had initially set out to solve. And yet I continued to push. I continued to fight. I planned and worked and searched and sang and cried, encountering failure upon failure upon failure. But I didn't let these failures rend my soul, tear me to pieces, break me in half, no. Instead, I fought. And eventually, I won. I conquered the darkness. Now, thinking back on this time of my life, I can't help but laugh ... To think that only a year ago I had believed that the entirety of my life's destiny rested upon a single disobedient slice of pigskin. And it was through these innumerable failures that I learned eventually to be fearless ... to be fearless to fight.

He concludes his speech with the following:

> As we step off the campus of Greenwich High School for the last time as students, we gain a new responsibility ... a responsibility to be fearless. And now as I stand upon this stage I urge you to take this responsibility in stride. I ask that you do not be afraid to be a dreamer.

To be an explorer. To be a fighter. To be united. I ask that you do not be afraid to innovate in developing solutions to our world's greatest problems, to fight against the forces of hatred, prejudice, injustice, and inequality, to launch our world into a better and brighter future than the one we already have.

For all of us, fear is but a false barrier standing in the way of progress, a demon standing in the way of enlightenment. In the words of Frank Herbert, "Fear is the mind-killer, the little-death that brings total obliteration. Where the fear has gone there will be nothing. Only I will remain." And so I urge you to conquer this fear. I urge you to stand up and be fearless. Because only by overcoming your fears may you come to change the world for the better. Thank you and congratulations, Class of 2017!

It's a powerful speech to anyone in the audience, but a poignant one to those who know William. Because despite his procrastination, his constant state of overload, his sloth-like gait, and his quiet manner, he has emerged, without a doubt, as a fearless human being. In his four years of high school, he hasn't let anyone or anything mar his pursuits. In trying to figure out who he is, he's not afraid to step outside of himself—to drink Soylent or wear an idiosyncratic jacket, to sing and dance in the high school musical.

But there's been no greater display of this than his time with Andy in room 932. It was with his science research where he took tremendous risks and the biggest of swings, never playing it safe or setting his sights any lower than changing the world.

It's not that he's never unconfident, hesitant, or self-doubting. At times, he is all of these things.

But one thing William Yin is not is fearful.

52

Andy

It's the night of Andy's annual Yankees outing with the science research kids.

Thursday, June 22, is a perfect, glistening evening for a ball game. The sweaty summer heat has mostly cooled off and there's a gentle breeze coursing through the stands. Yankee Stadium hums with its signature blend of anticipation and hardened grit, a kind of world-weary optimism beneath blinding lights and the grind of the subways.

Andy and about fifteen of his research kids have trekked in on the Metro-North train from Greenwich. Nothing trumps the need for food, so the gang first eats at the in-stadium Lobel's, where Andy has been coveting the prime rib sandwich. William and Verna have joined and they venture off on their own to get sushi, which the new Yankee Stadium, opened in 2009, actually serves.

At the end of the first inning, Andy and the research posse, which tonight includes his daughter, Sofia, her friend, and two Greenwich chemistry teachers, make their way to the bleachers, the cheapest and most raucous seats in the house, the section where beer and punches have been known to fly—and where a security guard is prominently stationed, ready in an instant to toss out the overly rowdy.

The kids are giddy. In a train ride, they're out from behind the hedges of Greenwich and in the heart of the Bronx. It's an assault on their senses and their sensibilities. They're seated beside people wearing gold chains with Yankee medallions the size of poker chips. They're among the impassioned fans who scream whenever possible, who jump up and fist-pump and chest-bump. They're not in golf-clap territory anymore. And tonight, the bleacher crowd is doing its part to uplift the bleacher glory.

Romano's wearing his Yankees cap. As he takes a seat, he says, "This bottle of water cost five dollars and twenty-five cents. It better cure cancer." But Romano's sticker shock at the bottled water is offset by his excitement at seeing a Nuts4Nuts stand outside the stadium. He's enamored with the hot candied nuts sold from carts on New York City street corners and he's already stoked to get some on the way out.

"Love the Nuts4Nuts!" he says. "The best!"

Rahul is sipping a soda he notes was $6.25 but mockingly points out, "But it comes in a souvenir cup!"

At one point, several of the kids head off in search of cotton candy. They come back taking *T. rex*–sized chomps out of the turquoise spun sugar, briefly looking like trolls with long, wispy blue beards before they gobble the candy into their mouths. Their lips, gums, and teeth turn blue, and the effect is especially pronounced on those with braces, their little metal brackets now framed in turquoise. They flash their smiles and you can't help but laugh.

The Yanks are playing the Los Angeles Angels, and after a sleepy first inning, the Bombers score four runs in the second, thanks to an Aaron Judge home run that electrifies the crowd. Everyone's up on their feet, whooping and cheering.

As for Andy, here he is, back home in the Bronx, about six miles from the lot where he slept many a hot night in the metal trailer earning his extra thirty dollars. Unlike his suburban troupe, he bears none of the wide-eyed big city glaze when he's here. He's back in his native environment and is as comfortable here as anywhere in the world.

Tonight he's rooting for his lifelong team, surrounded by his kids. There's no more science, no more long nights, no more experiments. The noise and fury of the year are rapidly fading, seeming like the distant roar heard miles away from the stadium after a home run soars over the top of its walls. It's a din that recedes and leaves an echo in your mind. If you close your eyes, you can summon it. And just as fast, you can let it go.

Ahead, there is nothing but uncertainty for Andy. He's not sure how long he'll stay at Greenwich High. And no matter what, he knows it's likely the beginning of the end. All of this unsettles his deep need for stability, for known contours and sure footing.

But for tonight, the ball game is it. From his posture to his gait to the muscles along his jawline, everything has eased. He keeps looking down row 11, checking on his kids, registering their happiness, and when he sees it, he sits back and sinks into his own.

A Record Year

I n the eleven years Andy Bramante has been teaching his science research class at Greenwich High School, 2016–17 was the most successful year yet in terms of wins. His students dominated the science fair circuit in unprecedented numbers.

- **Siemens Competition in Math, Science and Technology**
  Olivia Hallisey was named a semifinalist.

- **Google Science Fair**
  Shobhita Sundaram and William Yin were named regional finalists.

- **Regeneron Science Talent Search (STS)**
  Olivia Hallisey, Ethan Novek, Sanju Sathish, Derek Woo, Devyn Zaminski, and Madeleine Zhou were named semifinalists. They each won $2,000 with an additional $2,000 per semifinalist going to Andy's class.

Ethan Novek and Derek Woo were named finalists. Ethan placed eighth overall, winning $62,000. Derek won $27,000.

- **Connecticut STEM Fair (CT STEM)**

  Connor Li and Agustina Stefani won first place in their categories and $75 each. Olivia Hallisey and Shobhita Sundaram tied for first place in their category. Connor, Agustina, and Shobhita won all three of CT STEM Fair's Intel berths—a clean sweep for Greenwich High.

- **Connecticut Science and Engineering Fair (CSEF)**

  Luca Barcelo, Ethan Novek, Rahul Subramaniam, and Michelle Xiong won four of CSEF's seven Intel berths.

  Shobhita Sundaram won first place in life sciences and first place in mathematics.

  Michelle Xiong won first place in biotechnology and fifth place in physical sciences.

  Ethan Novek won first place in physical sciences and first place in environmental sciences.

  Manuel Carballo won second place in environmental sciences.

  Steven Ma won second place in mathematics.

  Shobhita Sundaram won second place in computer sciences.

  Michelle Woo won second place in environmental/renewable energies.

  Manuel Lopez won third place in biotechnology.

  Agustina Stefani won third place in physical sciences.

  Sophia Chow won fourth place in biotechnology.

- **Connecticut Junior Sciences and Humanities Symposium (JSHS)**

  Rahul Subramaniam and Luca Barcelo were selected to present posters of their work.

  Connor Li, Shobhita Sundaram, and William Yin were selected to deliver oral presentations of their projects.

Shobhita placed first and won $2,000.

William won second place and $1,500. As part of their wins, they were sent to the National JSHS competition, where Shobhita won first place in the category of medicine and $12,000.

- **Intel International Science and Engineering Fair (Intel ISEF)**
Rahul Subramaniam won a first-place award and $3,000 in the category of microbiology. He also won best in category and $5,000, with an additional $1,000 going to Andy's program, as well as the Intel Indo-U.S. Science and Technology Forum Scientific and Cultural Visit to India Award.
Luca Barcelo won a second-place Grand Award and $1,500 in environmental engineering, as well as a tuition scholarship to the University of Arizona.
Shobhita Sundaram won a fourth-place Grand Award and $500 in computational biology, a first-place specialty mathematics award of $1,000 from the National Security Agency Research Directorate, and an honorable mention from the American Statistical Association.

- **International Sustainable World (Engineering Energy Environment) Project (ISWEEEP)**
Rahul Subramaniam and William Yin won Gold medals.
Sophia Chow, Bennett Hawley, and Sanju Sathish won Silver medals.
Dante Grace Minichetti won a Bronze award.

- **Norwalk Science Fair**
Collin Marino, Alex Kosyakov, and Michelle Woo placed first through third, respectively, winning $300 to $100.
Joe Konno won third place, honorable mention.
Romano Orlando won sixth place, honorable mention.

- **Phenomenon: Science Innovation Fair**
Rahul Subramaniam won first place.

Dante Grace Minichetti won second place.

Michelle Woo won the People's Choice Award.

* **GENIUS Olympiad**

Takema Kajita, Alex Kosyakov, and Michelle Woo all won silver medals and Kindle Fire tablets.

* **Davidson Fellows Scholarship**

William Yin received a $25,000 scholarship.

Life Beyond GHS

Here's where the GHS Class of 2017 science research students are going to college:

Jacob Back:	University of Rochester
Manuel Carballo:	Cornell
Saralexi Chacon:	Sacred Heart University
William Chen:	University of Pennsylvania
Margaret Cirino:	University of Southern California
Henry Dowling:	Harvard
Alessio Fikre:	Harvard
Olivia Hallisey:	Stanford
Takema Kajita:	Cornell
Connor Li:	Cornell
Ethan Novek:	Accepted early to Yale, deferred admission for a year
Christo Popham:	Princeton
Edouard Quiroga:	University of Connecticut
Adam Roitman:	Duke

Sanju Sathish: Stanford

"Danny Slate": Harvard

Agustina Stefani: University of Virginia

Derek Woo: Harvard

William Yin: Stanford

Devyn Zaminski: Vanderbilt

Madeleine Zhou: University of Chicago

And finally, here is where the members of Andy Bramante's class of 2018 will continue their educations:

Alex Araki: University of Wisconsin

Luca Barcelo: Columbia

Sophia Chow: University of California, Berkeley

Wesley Heim: Dartmouth

Noah Kim: Vanderbilt

Manuel Lopez: Brown

Steven Ma: Yale

Dante Grace Minichetti: Wellesley

Romano Orlando: University of Southern California

Emily Philippides: Princeton

Amit Ramachandran: Stanford

Henry Shi: Stony Brook University

Haley Stober: University of California, Berkeley

Shobhita Sundaram: MIT

Michelle Woo: Princeton

Michelle Xiong: University of Pennsylvania

Acknowledgments

O n a cold fall day in 2015, I nervously went to lunch with Andy and Tommasina Bramante, saying that I had a project I wanted to discuss with them. It wasn't until the very end of our meal, after the plates had been cleared, that I finally blurted out to Andy, "I want to write a book about you."

We'd met a few weeks prior when I went to his lab to produce a piece for the *CBS Evening News* about Olivia Hallisey after she won the Google Science Fair. When I stepped into Andy's classroom, I saw glimpses of the extraordinary—a rare, wonderful teacher of bafflingly smart, driven kids who took part in the heady world of competitive science. I knew that day I wanted to write a book about Andy and his students—and even quit my job at CBS News to do so.

So to Andy—thank you for letting me into your classroom and your life. The year spent in room 932 was the most satisfying of my career. Your humor and kindness infused the process with more fun than I thought possible. So many Andy-isms are now part of my lexicon. And as for what you do, the subtitle of this book says it all. You truly change your kids' lives, and in allowing me along for the ride last year, you changed mine.

To the families: the Yins, Slates, Chows, Orlandos, Halliseys, and Noveks—I am so grateful for your time and generosity in allowing me to chronicle a year in the lives of your kids. And to William, Danny, Sophia, Romano, Olivia, and Ethan—what a tremendous privilege to get to know all of you. I learned so much from each of you; you have my deepest admiration.

To all the other science research students during the 2016–17 school year—I loved spending time with you. Thank you for every single minute you gave to me.

There are many other parents and people who provided their time and insights, including Charles Woo and Ryeo-Jin Kang, Chitra Sundaram, Cindy and Juerg Heim, Sandy Minichetti, Anthony and Sally Mann, Reed Newberry, and Ray Hamilton.

To those who work tirelessly to keep the science fair circuit thriving each year and who spoke to me: Sandra and Wynn Müller, Robert Wisner, Joy Erickson, Zia Mannan, and Michele Glidden. I'm grateful to Potoula Gjidija for facilitating my interview with Dr. George Yancopoulos of Regeneron. And George—thank you for taking some time from running your company to talk to me.

I'm thankful to several people at Greenwich High School. Dr. Chris Winters allowed me to spend the year in Andy's science research class and gave invaluable insights on the GHS student body. Guidance counselor Suzanne Patti kindly took time out of her packed days to speak to me. Chemistry teachers Shirley Barban and Cindy Vartuli provided endless kindness and humor.

To Ted Ohls and Harold Williams, my Greenwich parking gods—you rose to the occasion when no one else would. You have my deepest thanks.

To my agent, David Black, thank you for sticking with me through the proposal process, and for always saying, "This is going to be a terrific book." You should know that during our very first phone conversation nothing registered because all I could think, on a constant loop, was "Holy crap. I'm talking to David Black."

To the wonderful people at Penguin Random House: Gina Centrello, Jennifer Hershey, Kara Welsh, Kim Hovey, Matthew Martin, Susan Corcoran, Michelle Jasmine, Allyson Lord, Kristin Fassler, Debbie Aroff, Danielle Siess, Steve Messina, Amy Ryan, Rachel Ake, Joe Perez, Paolo Pepe, and Emily Hartley. Thank you for your outsize support of my book and for making the process so rewarding.

To my editor, Brendan Vaughan—after 125,000 words, I leave you with just four: you are the best. If I could, I'd fly a banner-tow airplane over your house daily with those words. I'd insert them into your fortune cookies and swirl them into your latte foam. You're a supernova of an editor who made this process better than I could have ever imagined. More important, you're a fantastic human being. I am so grateful for your sharp eye, humor, warmth, attentiveness, friendship (I'm wrapping this up because I know you already think there are too many descriptors here), and curatorial genius when it comes to television. Thank you, a thousand times, thank you. I shall forever raise a thirty-two-ounce beaker mug in your honor while listening to Aaron Copland.

I'm grateful beyond words to Kira Henehan and Hilary Elkin, the indomitable researchers who signed on to fact-check the book when I was nine months pregnant. You came on board when I needed nothing more than to feel as if the book was in great hands. You did an amazing job. This is a better book because of your work.

I owe so much to family and friends who have supported and encouraged me during the writing of this book, and beyond.

To my parents, Robert and Nancy Tesoriero, who never once restricted my dreams and aspirations—one of the greatest gifts a parent can give. Also, one of the best services you ever did for me was not applying any pressure on me in school or in life! Thank you for the constant love you give to your granddaughters. And finally, thank you for being the kind of people who were literally willing to reach across the world to become my parents. I hope I've made you proud. Love you.

To all the Cranes—thank you for welcoming me to the family and being such wonderful grandparents. Mary—mahalo for your support, enthusiasm, and always cheerful outlook. Roy and Sheila—what a pleasure it was to write this book on the MacBook Air you gave me! Best gift ever.

Other relatives to whom I'm grateful: Lois and Joe Mahonchak and their kids, Beatrice Marin, Kaz Tokunaga, Ira and Judy Kiyomura, Dana Kiyomura and Richard Tilles, and my late grandfather, Angelo Tesoriero, whose memory I hold dear.

To the friends who provided love, encouragement, assistance, guidance, humor, meals, and lodging: Kathy Mah and Matt Moore, Scott Hensley and Lisa Sanders, Sarah Tombaugh and Doug Vanderpool, Julian and Jinan O'Connor, Heather and Joe Spinelli, Vanessa Fuhrmans and Troy McMullen, Samantha Amato and Michael Rotkowitz, Melissa and John Caruso-Scott, Glenn and Julie Morey, Lara Allison and Matt Warren, Natasha Henley and Mark Stevenson, Julia and Kurt Frey, Rosa Park and Jason Ma, Matt Nelson and Tim Hayne, Deeksha Gaur and Joe Slaughter, Xenia von Lilien and Chris Brockmeyer, Tara and Dave Cangello, Delmi Gonzalez, Susannah Meadows, Ruth Ann Stanley, Jihee Woo, Jenna Shnayerson, Tracy van Straaten, Jenny Turner Hall, Colin Cowley, Radiah Harper, and Jared Blake.

Susan Pelzer, there's not enough space here to express the depths of my love and gratitude. No friend read more drafts of my proposal or more incremental updates throughout the whole process. And thank you for giving me the gift of writing in Geoff's study in July 2016. That was one of my most fruitful days ever, owing to wanting to make you and Geoff proud.

Sheryl Hastalis and Nina Lane—without the Mirsky sisters, there would be no journalism career for me. It all started when you kindly let me live in Nina's condo during my first magazine internship. Sheryl, you remain my most loyal reader. You always saw me first and foremost as a writer, and it truly helped steady my course when I felt adrift. Love you both. (And Michael and Jim, too!)

Friends, colleagues, and friends of friends who were so helpful when I was seeking representation: Michael Shnayerson, Claudia Kalb, Cassi Feldman, Marcus Brauchli, Joshua Prager, Christine Kenneally, Chris Rhoads, Chuck Wilson, and Judith Mattson.

To the CBS News med unit: Dr. Jon LaPook, Kevin Finnegan, Amy Birnbaum, Dana Carullo, and Susie Schackman—thanks for sending me out into the world with so much love. I'm beyond grateful for our eight years together. I know you miss my typing and listening while I record my outgoing voicemail greeting.

To Jim Axelrod—how many people can you say you've spent time with in a moving truck in the Bronx being fumigated for bedbugs? That would be one and that would be me. All for the glory of network television. What neither of us knew that day was that you would be the sine qua non of this book. And I was enormously bolstered by every gem you passed along from the experience of writing your own book. Thank you for sending me a note at the very start of this process that said, "I believe in you." Your belief was paramount, as is your friendship.

To Rachel Packman Chou for generously providing me with a magical place to write in the fall of 2017 when I was fast approaching deadline. It was an act of tremendous heart and kindness that I'll never forget. And I so appreciated your video tour of the house!

Thank you to every single friend who responded to my Facebook plea for a getaway during the final stretch. I am still awed by all of your offers.

I would be remiss if I didn't include my own life-changing teacher, Monnie Murtha, the first person who identified me as a writer while I was in first grade at Riley Avenue Elementary School. You're there with Andy Bramante in the pantheon of amazing teachers. Love you, MM.

To my daughters, G + E. For so many reasons, this book is for you and our family. G—you have the purest and most beautiful heart of anyone I've ever known. My daily trek to Greenwich was made easier knowing I'd come home to you and your smile. (And I love that since infancy you've been a book person.) E—you can one day tell people that you've been to Intel ISEF, a junior prom, and a high school graduation—all before birth. Being pregnant with you during the writing of this book kept me going because I knew I had you to look forward to. To you both, while I am tremendously proud of this book, I will always be proudest of being your mom.

And, finally, to JH—you are the dream who makes all my dreams possible. For how much I love you, there are no words.

Image Credits

INSERT PAGE ONE

Sophia and Andy in lab: courtesy of the author

Romano in lab: courtesy of the author

Ethan in lab: courtesy of Ethan Novek

INSERT PAGE TWO

William: courtesy of William Yin

William and Verna: courtesy of the author

William in band: courtesy of the author

William prom: courtesy of William Yin

INSERT PAGE THREE

Instagram: courtesy of the author

Olivia and father: courtesy of Andy Bramante

Sophia in lab: courtesy of the author

Hands in lab: courtesy of the author

INSERT PAGE FOUR

Andy and family: courtesy of Andy Bramante

Dog: courtesy of Andy Bramante

INSERT PAGE FIVE

Boys and McCain: courtesy of Charles Woo

Michelle in lab: courtesy of the author

Girls at competition: courtesy of the author

INSERT PAGE SIX

Ethan on camera: courtesy of Andy Bramante

Sophia and dad: courtesy of Gregg Chow

Andy and students: courtesy of William Yin

INSERT PAGE SEVEN

Romano and Sophia: courtesy of the author

Prom group: courtesy of Andy Bramante

INSERT PAGE EIGHT

Finalists and food: courtesy of Ethan Novek

Baseball game: courtesy of the author

Andy and William: courtesy of the author

ABOUT THE AUTHOR

HEATHER WON TESORIERO is an Emmy-winning former producer for CBS News and has been a reporter at the *Wall Street Journal, Time,* and *Newsweek.* A Korean adoptee who was discovered on a doorstep when she was just a few days old, she grew up on the eastern end of Long Island and now lives in New York City with her husband and two young daughters.

ABOUT THE TYPE

This book was set in Albertina, a typeface created by Dutch calligrapher and designer Chris Brand (1921–98). Brand's original drawings, based on calligraphic principles, were modified considerably to conform to the technological limitations of typesetting in the early 1960s. The development of digital technology later allowed Frank E. Blokland (b. 1959) of the Dutch Type Library to restore the typeface to its creator's original intentions.